MICROBIOLOGY REVIEW

By

CHARLES W. KIM, Ph.D.
Associate Professor
Department of Microbiology
Health Sciences Center
State University of New York
Stony Brook, New York

FIFTH EDITION

**1800 MULTIPLE
CHOICE QUESTIONS
AND ANSWERS
—
COMPLETELY
REFERENCED**

BASIC SCIENCE REVIEW SERIES

MEDICAL EXAMINATION PUBLISHING COMPANY, INC.
65-36 Fresh Meadow Lane
Flushing, N.Y. 11365

Copyright © 1973 by the Medical Examination Publishing Co., Inc.

Copyright ©.1973 by the Medical Examination Publishing Co., Inc.

All rights reserved. No part of this publication may be reproduced in any form or by any means, electronic or mechanical, including photocopy, without permission in writing from the publisher.

Library of Congress
Catolog Card Number
62-14605

ISBN 0-87488-203-6

April 1973

PRINTED IN THE UNITED STATES OF AMERICA

PREFACE

There are no short cuts to the process of learning! The material on the following pages is not designed to provide the reader with an opportunity to escape from learning basic information that can only be obtained from textbooks. The purpose of this text is to encourage the reader to detect areas of weakness in his understanding of subject matter so that he may return to his texts for a more comprehensive review of the subject. Crossword puzzles are not designed to teach basic English; examination review books do not teach a basic understanding of the subject. However, the following pages will provide an interesting challenge to the student as well as an opportunity to improve his skills with multiple-choice examinations.

The completely referenced form of the question material is provided to facilitate the reader in checking his areas of weakness. Do not compromise by simply looking up the correct answers — the mature approach is to return to your textbook. The author has provided you with a time-saving device in rapidly finding a source of information. While testing your memory be aware of the textbook as being the ultimate source of your fund of knowledge. References cited in the individual questions are in the back of the book.

MICROBIOLOGY REVIEW

FIFTH EDITION

TABLE OF CONTENTS

SECTION

I	BACTERIAL PHYSIOLOGY.. Questions 1-240	7
II	IMMUNOLOGY .... Questions 241-460	35
III	BACTERIOLOGY....... Questions 461-910	58
IV	VIROLOGY .... Questions 911-1300	104
V	MEDICAL MYCOLOGY.... Questions 1301-1450	142
VI	MEDICAL PARASITOLOGY . Questions 1451-1650	156
VII	GENERAL REVIEW OF INFECTIOUS DISEASES ... Questions 1651-1800	176
REFERENCES		193
ANSWER KEY		194

SECTION I - BACTERIAL PHYSIOLOGY

FOR EACH OF THE FOLLOWING MULTIPLE CHOICE QUESTIONS, CHOOSE THE ONE MOST APPROPRIATE ANSWER:

1. CONCERNING BACTERIAL CHROMOSOME:
   A. It is composed of DNA
   B. It contains no histones
   C. Many of the DNA molecules extracted are circular
   D. The DNA accounts for about 2 to 3 per cent of the dry weight of the cell
   E. All of the above
   Ref. 6 - p. 106

2. THE MOST PRIMITIVE MODE OF GENE TRANSFER OCCURS BY:
   A. Transformation
   B. Transduction
   C. Conjugation
   D. Cell fusion
   E. All of the above
   Ref. 6 - p. 136

3. THE STAGE OF GROWTH IN WHICH THERE IS A SLOW LOSS OF CELLS THROUGH DEATH, WHICH IS JUST BALANCED BY THE FORMATION OF NEW CELLS THROUGH GROWTH AND DIVISION:
   A. Lag
   B. Acceleration
   C. Exponential
   D. Maximum stationary
   E. Decline
   Ref. 15 - p. 80

4. A GENE SEQUENCE UNDER THE COORDINATED CONTROL OF A SINGLE OPERATOR IS CALLED:
   A. Repressor
   B. Operon
   C. Regulator gene
   D. Structural gene
   E. None of the above
   Ref. 15 - p. 46

5. THE FOLLOWING IS REQUIRED BY LIVING CELLS AS A COMPONENT OF ATP, OF NUCLEI ACIDS, AND OF SUCH COENZYMES AS NAD, NADP, AND FLAVINS:
   A. Organic sulfur
   B. Inorganic phosphate
   C. Magnesium ion
   D. Ferrous ion
   E. Potassium ion
   Ref. 15 - p. 74

6. THE BACTERIAL CELL MEMBRANE CAN BE DEMONSTRATED BY THE FOLLOWING METHOD:
   A. Plasmolysis
   B. Staining
   C. Isolation
   D. Ultrathin sections
   E. All of the above
   Ref. 15 - pp. 9-10

7. FACULTATIVE ANAEROBES:
   A. Are unable to grow in the presence of molecular oxygen
   B. Require an excess of oxygen
   C. Require no oxygen
   D. Respire with equal facility in the presence or absence of molecular oxygen
   E. Are not pathogenic
   Ref. 2 - p. 111

8. OBLIGATE ANAEROBIC BACTERIA:
   A. Oxygen is not toxic to them
   B. Vegetative cells die when exposed to oxygen
   C. All contain catalase
   D. Spores also die upon exposure to air
   E. Contain large amounts of iron-porphyrin respiratory enzymes
   Ref. 2 - p. 111

SECTION I - BACTERIAL PHYSIOLOGY

9. IN BACTERIA:
   A. ADP and ATP have the same number of energy-rich bonds
   B. ATP is not generated in anaerobic oxidation
   C. ATP is generated in aerobic and anaerobic oxidations
   D. ADP has two bonds which are active in energy transfer
   E. Energy is not stored in ATP   Ref. 2 - p. 114

10. IN MOST SPECIES OF HETEROTROPHIC BACTERIA:
    A. Protein serves as the chief source of energy
    B. Fats serve as the chief source of energy
    C. Carbohydrates serve as the chief source of energy
    D. Carbohydrates have only one function
    E. Carbohydrates do not serve as a source of carbon
       Ref. 2 - p. 114

11. BACTERIA BREAK DOWN COMPLEX SUGARS BY:
    A. Hydrolysis   D. All of the above
    B. Phosphorolysis   E. None of the above
    C. Transglycosidation   Ref. 2 - p. 115

12. AUTOTROPHIC BACTERIA:
    A. Do not require organic material for growth
    B. Do not require a nitrogen source for growth
    C. Are always photosynthetic
    D. Are never chemosynthetic
    E. Do not require essential inorganic ions
       Ref. 2 - p. 132

13. HETEROTROPHIC BACTERIA:
    A. Require an oxidizable organic compound as an energy source for growth
    B. Do not require a nitrogen source for growth
    C. Are never photosynthetic
    D. Are never chemosynthetic
    E. Do not utilize CO_2 Ref. 2 - p. 132

14. IN TRANSAMINATION:
    A. Amino group of one amino acid is transferred to any position of a keto acid
    B. A new amino acid is synthesized
    C. A new keto acid is not synthesized
    D. Transaminases are not essential
    E. Glutamic and alpha-ketoglutaric acids are never involved
       Ref. 2 - pp. 145-146

15. AMINO ACID RACEMASE:
    A. Converts only the L-amino acid
    B. Converts only the D-amino acid
    C. Converts either D- or L-amino acids into a racemic DL-mixture
    D. Does not act as a catalyst
    E. Is not present in bacteria   Ref. 2 - p. 146

16. IN PROTEIN SYNTHESIS:
    A. Peptide bonds are unimportant
    B. The energy is not supplied by a nucleoside triphosphate
    C. There is at least one transfer RNA for each amino acid
    D. All amino acids are not linked to their transfer RNA's through esterification
    E. The reaction proceeds actively in the ribosomes but not in the cytoplasmic membrane   Ref. 2- p. 166

SECTION I - BACTERIAL PHYSIOLOGY

17. A NUCLEOTIDE IS:
    A. Not a naturally occurring nitrogenous substance
    B. Not composed of either ribose or deoxyribose
    C. Composed of a purine or pyrimidine base
    D. Not composed of a molecule of phosphoric acid
    E. Composed of glucose   Ref. 2 - p. 158

18. NUCLEIC ACIDS:
    A. Are found only in the cell nucleus
    B. Are found only in the cell cytoplasm
    C. May occur in nature in combination with proteins as nucleoproteins
    D. Do not vary from cell to cell
    E. Bacteria have small amounts of nucleic acid as compared to higher forms   Ref. 2 - p. 159

19. IN THE BREAKDOWN OF NUCLEIC ACIDS:
    A. All cells have enzymes for degrading nucleic acids to smaller fragments
    B. Nucleodepolymerases attack intact nucleic acids
    C. Deoxyribonucleases (DNase) hydrolyze phosphate ester linkages in DNA
    D. Ribonucleases (RNase) appear to be of a single type
    E. All of the above   Ref. 2 - p. 159

20. PURINES AND PYRIMIDINES:
    A. Interconversions may occur only at the level of the free base
    B. Are synthesized by completely independent pathways
    C. Are not structurally related
    D. Unlike the synthesis of amino acid, folic acid coenzymes are not involved
    E. Cannot be further broken down by bacteria
          Ref. 2 - p. 160

21. CONCERNING DNA:
    A. The DNA molecules can make more molecules like itself but cannot cause the formation of RNA molecules whose nucleotide sequences carry some of the information possessed by the DNA
    B. Each DNA molecule consists of two polynucleotide strands coiled about a common axis
    C. Hydrogen bonds are formed anywhere along the chain
    D. In a single strand of DNA only one sequence of nucleotides is possible
    E. Each strand in a molecule of DNA is not the complement of the other
          Ref. 2 - p. 164

22. A GROWTH FACTOR MAY BE BEST DEFINED AS A (AN):
    A. Vitamin
    B. Nucleoprotein
    C. Protein
    D. Nucleotide
    E. Organic compound which a bacterium needs but cannot synthesize
          Ref. 2 - p. 168

SECTION I - BACTERIAL PHYSIOLOGY

23. THE B-COMPLEX VITAMINS ARE:
    A. Low molecular weight
    B. Water-soluble
    C. Organic compounds
    D. Universal constituents of all living cells
    E. All of the above   Ref. 2 - p. 168

24. NICOTINIC ACID IS:
    A. Not required for bacterial growth
    B. Needed for the synthesis of NAD and NADP
    C. Vitamin B_6
    D. Not needed for the synthesis of coenzymes which function in hydrogen transport
    E. Utilized by all species of Hemophilus   Ref. 2 - p. 169

25. VITAMIN B_6:
    A. Is a specific chemical
    B. Is not a growth factor
    C. Is a good example of a growth factor which is highly stimulatory, but not absolutely required
    D. Is pantothenic acid
    E. Is para-aminobenzoic acid   Ref. 2 - p. 171

26. IN GENERAL, GRAM-NEGATIVE BACTERIA:
    A. Synthesize more amino acids than gram-positive bacteria
    B. Synthesize less amino acids than gram-positive bacteria
    C. Synthesize the same amount of amino acids as gram-positive bacteria
    D. Cannot synthesize many of their amino acids
    E. Require preformed amino acids   Ref. 2 - p. 174

27. VITAMIN B_{12}:
    A. Contains an atom of cobalt bound to the porphyrin-like structure
    B. Is not required by any bacteria
    C. Has no metabolic function
    D. Has a simple chemical formula
    E. None of the above   Ref. 2 - p. 173

28. BREAKDOWN OF AMINO ACIDS:
    A. Is accomplished only by decarboxylation
    B. Is accomplished only by deamination
    C. Involves constitutive enzymes
    D. Involves adaptive enzymes which catalyze decarboxylation and deamination
    E. None of the above   Ref. 2 - p. 143

29. AMINO ACID:
    A. Is an important energy source for most bacteria
    B. Its metabolism is less difficult to visualize than carbohydrate metabolism
    C. Both the degradative and synthetic phases of amino acid metabolism are closely integrated with the metabolism of carbohydrates
    D. Does not occur in bacteria in the free state
    E. Occurs in bacteria only in peptides   Ref. 2 - p. 142

SECTION I - BACTERIAL PHYSIOLOGY

30. THE STICKLAND REACTION:
    A. Does not require specific enzymes
    B. Is a reaction in which one amino acid is oxidized, another reduced
    C. Any hydrogen donor does not give a Stickland reaction with any hydrogen acceptor
    D. Is not capable of serving as the chief energy source for the proteolytic clostridia
    E. Does not proceed at a rate comparable to the aerobic oxidation of carbohydrate in fermenters   Ref. 2 - p. 145

31. MOST BACTERIA ARE:
    A. Psychrophiles
    B. Mesophiles
    C. Cryophiles
    D. Thermophiles
    E. None of the above
    Ref. 12 - p. 135

32. THE WORK OF AVERY AND GOEBEL ON BACTERIAL TRANSFORMATION WAS ACCOMPLISHED WITH:
    A. <u>Bacillus subtilis</u>
    B. Streptococcus
    C. Pneumococcus
    D. <u>Escherichia coli</u>
    E. <u>Hemophilus influenzae</u>
    Ref. 12 - p. 184

33. THE TRANSFER OF GENETIC INFORMATION FROM DNA TO RNA IS CALLED:
    A. Transduction
    B. Transcription
    C. Transformation
    D. Recombination
    E. Conjugation
    Ref. 12 - p. 227

34. MESSENGER RNA HAD PREVIOUSLY BEEN OVERLOOKED IN CHEMICAL ANALYSIS BECAUSE:
    A. Usually it does not remain complexed with the ribosomes in the course of cell extraction
    B. It is ordinarily a large fraction of the total RNA of the cell
    C. It is homogeneous in molecular weight
    D. It is especially susceptible to destruction by ribonuclease
    E. It forms a visible peak in the ultracentrifuge
    Ref. 12 - p. 273

35. IN ENDPRODUCT INHIBITION OF ENZYME FUNCTION:
    A. Only the initial enzyme of a sequence is affected
    B. All the enzymes of a sequence are affected
    C. The mechanism is influenced by same mutations as in repression
    D. All of the above
    E. None of the above   Ref. 12 - p. 283

ANSWER THE FOLLOWING QUESTIONS (T)RUE OR (F)ALSE:

36. The chemical substance of chromosome which is responsible for both gene replication and gene function in viruses is either DNA or RNA.
    Ref. 15 - p. 32

37. In a bacterial cell, several different genetic (DNA) structures cannot be present and replicating independently at the same time.
    Ref. 15 - p. 34

SECTION I - BACTERIAL PHYSIOLOGY

38. The set of genetic determinants carried by a cell is called its genotype while the observable properties of the cell are called its phenotype.
 Ref. 15 - p. 37

39. The recombinant chromosome formed as a result of transformation consists of double-stranded DNA of the recipient in which a short region of one of the two strands has been replaced by a strand of donor DNA.
 Ref. 15 - p. 40

40. Cells of E. coli K12 which carry F (the sex factor) are called F^-; Those which have lost it are called F^+.
 Ref. 15 - p. 44

41. When a suspension of Hfr (high frequency of recombination) cells is mixed with an excess of F^- cells, every Hfr cell in the culture transfers chromosomal DNA to an F^- cell. Ref. - pp. 44-45

42. In addition to the Embden-Meyerhof and the direct oxidative pathways, many microorganisms break down hexose via the Entner-Doudoroff pathway. Ref. 15 - pp. 63-64

43. Most microorganisms cannot use NH_3 as the sole nitrogen source; however, most of them can fix atmospheric nitrogen (N_2) and convert it to NH_3 in the cell. Ref. 15 - pp. 73-74

44. For a microbial cell, death means the irreversible loss of the ability to reproduce. Ref. 15 - p. 81

45. The measurement of microbial death usually refers to the death of a population rather than the death of an individual cell.
 Ref. 15 - p. 82

ANSWER THE FOLLOWING QUESTIONS BY USING THE KEY OUTLINED BELOW:
A. If 1 and 4 are correct
B. If 2 and 3 are correct
C. If 1, 2 and 3 are correct
D. If all are correct
E. If all are incorrect
F. If some combination other than the above is correct

46. THE PERIPHERAL PART OF THE BACTERIAL CELL MAY BE DIVIDED INTO THE:
 1. Capsule
 2. Cell wall
 3. Cytoplasmic membrane
 4. Endoplasmic membrane Ref. 5 - pp. 43-44

47. CONCERNING BACTERIAL CAPSULES:
 1. Bacteria may lose their capsules through mutations
 2. Such loss of capsule is reflected in altered appearance of the colonies
 3. Encapsulated bacteria form more or less mucoid colonies
 4. Bacteria without capsules form relatively dry colonies with a smooth or rough surface Ref. 5 - p. 45

SECTION I - BACTERIAL PHYSIOLOGY

48. THE BACTERIAL ENDOSPORE:
    1. The spore core appears without much structural detail in electron micrographs
    2. The core is surrounded by a single membrane
    3. Inside this membrane is the cortex
    4. The content of dipicolinic acid is high   Ref. 5 - p. 61

49. THE BACTERIAL RNA:
    1. The so-called soluble RNA has a low molecular weight and is not bound to protein
    2. This soluble RNA is also called transfer RNA
    3. The messenger RNA carries genetic information from DNA to the ribosomes
    4. The messenger RNA is metabolically unstable and has a molecular weight of 250,000 or more   Ref. 5 - p. 56

50. GENETIC TRANSFORMATION OF PNEUMOCOCCI:
    1. The extraction of DNA from culture of donor pneumococci and its application to recipient pneumococci result in the transformation of some of the recipient organisms
    2. The latter does not acquire a differentiating property of the donor
    3. Cells which have been transformed for one property such as penicillin resistance are always transformed at the same time for another property such as capsulation
    4. The number of cells transformed is proportional to the amount of DNA applied to a saturation level of up to 10 per cent of all of the treated cells   Ref. 5 - p. 74

51. TRANSDUCTION:
    1. This was first demonstrated in S. typhimurium
    2. It involves the transfer of bits of genetic material from bacterium carried by a virus
    3. Following infection with temperate phages, a variable proportion of the infected cells will respond either by lysing or by being latently infected
    4. The gross genetic manifestations are quite different from those in transformation   Ref. 5 - pp. 75-76

52. BACTERIAL CONJUGATION EXPERIMENTS HAVE SHOWN THAT:
    1. Bacteria are haploid
    2. Bacteria are diploid
    3. They form a complete diploid heterozygote with a high frequency
    4. They form a complete diploid heterozygote with a low frequency
           Ref. 5 - p. 79

53. ENZYME REPRESSION:
    1. Operates at the genetic level
    2. It prevents transfer of genetic information
    3. Specific repressors are formed, presumably by way of specific messenger RNA
    4. The repressor interacts with the effector which would be the end-product metabolite or a product formed from the metabolite
           Ref. 5 - p. 117

SECTION I - BACTERIAL PHYSIOLOGY

54. THE FOLLOWING DRUGS ARE ALL BELIEVED TO ACT, AT LEAST IN PART, THROUGH INHIBITION OF PROTEIN BIOSYNTHESIS:
    1. Chloramphenicol
    2. Puromycin
    3. Streptomycin
    4. Tetracyclines   Ref. 5 - p. 128

55. THE ACTINOMYCINS:
    1. Have a useful role in therapy of some tumors
    2. Inhibit growth of DNA viruses
    3. Their primary action is due to complex formation with DNA
    4. They also bind to RNA   Ref. 5 - p. 136

56. CELL WALL:
    1. Is thinner in gram-negative bacteria than in gram-positive bacteria
    2. Is thinner in gram-positive bacteria than in gram-negative bacteria
    3. Makes up about 50 percent of the dry weight of the cell
    4. Is permeable to molecules as large as nucleotides
                                        Ref. 2 - p. 50

57. COCCI MAY BE CLASSIFIED AS:
    1. Micrococcus if they appear singly and scattered at random
    2. Diplococcus if they tend to occur in pairs
    3. Streptococcus if they tend to occur in chains
    4. Staphylococcus if they appear in irregular, grape-like clusters
                                        Ref. 2 - pp. 39-40

58. THE BACTERIAL FLAGELLUM:
    1. Is not used for locomotion but for obtaining food
    2. Is more common among the bacilli
    3. May be attached at either poles of the cells or may be distributed over the cell surface
    4. Cannot be antigenically distinguished from that of the cell proper
                                        Ref. 2 - p. 44

59. FLAGELLA:
    1. Have a characteristic amino acid composition
    2. Cannot be separated easily from the bacterial cell
    3. The motion of flagella is not a hydrodynamic phenomenon
    4. Probably have their energy source in ATP
                                        Ref. 2 - pp. 44-45

60. THE CELL WALL OF BACTERIA:
    1. Has not been proven to exist
    2. May be freed of intracellular material by mechanical disintegration of the cells
    3. Consists of peptide, polysaccharide and lipid
    4. Does not adsorb bacteriophage, but is digested by trypsin
                                        Ref. 2 - pp. 50-51

61. SPORE FORMATION:
    1. Occurs in response to adverse environmental conditions
    2. May occur centrally, subterminally or terminally
    3. Initially starts with the assembling or local concentration of material, presumably nucleoprotein
    4. Is a method of reproduction, because more than one spore may be formed in each cell   Ref. 2 - pp. 54, 55

SECTION I - BACTERIAL PHYSIOLOGY

62. THE MODE OF ACTION OF ANTIBACTERIAL AGENTS IN WHICH THERE IS INTERFERENCE WITH ONE OR A FEW SPECIFIC ENZYMATIC REACTIONS:
    1. Protein coagulation
    2. Disruption of cell membrane or wall
    3. Removal of free sulfhydryl groups
    4. Chemical antagonism
    Ref. 15 - p. 83

63. THE LAG PHASE:
    1. Is divisible into an apparent and a true lag phase
    2. Is dependent on the size of the inoculum within limits
    3. May be prolonged by growth inhibiting substances
    4. Directly follows the exponential growth phase
    Ref. 2 - p. 84

64. SUCCINIC DEHYDROGENASE CONSISTS OF AT LEAST THE FOLLOWING COMPONENTS:
    1. An iron-flavoprotein which oxidizes succinic acid to fumaric acid
    2. Slater's factor
    3. Cytochrome c-linking factor
    4. Cytochrome b
    Ref. 2 - p. 108

65. ENERGY SOURCES FOR BACTERIA INCLUDE:
    1. Oxidation of inorganic compounds
    2. Utilization of the energy of visible light
    3. Oxidation of organic compounds
    4. Absorption of heat
    Ref. 2 - p. 107

66. DISCRETE NUCLEAR BODIES IN BACTERIA CAN BE RECOGNIZED UNDER THE LIGHT MICROSCOPE THROUGH:
    1. Use of a DNA-specific stain
    2. Selective hydrolysis of the RNA by HCl or RNase, followed by staining with a basic dye
    3. Suspension of live cells in a medium of the same refractive index as the cytoplasm, which permits the nuclei to be recognized under phase contrast as areas of different refractility
    4. Suspension of dead cells in a medium of the same refractive index as the cytoplasm
    Ref. 12 - p. 39

67. SINCE BACTERIA LACK DISCRETE STRUCTURES FOUND IN THE NUCLEI OF EUCARYOTIC CELLS, THE NUCLEAR REGION IN BACTERIA IS REFERRED TO BY SOME AS THE:
    1. Nucleoid
    2. Nuclear body
    3. Chromatin body
    4. Nuclear equivalent
    Ref. 12 - p. 40

68. BACTERIAL SPORES:
    1. Are unusually dehydrated
    2. Are highly refractile
    3. Do not take ordinary stains
    4. Are sensitive to disinfectants
    Ref. 12 - p. 153

69. IN ANAEROBIC RESPIRATION THE:
    1. Final hydrogen acceptor is molecular oxygen (O_2)
    2. Final hydrogen acceptor is an inorganic compound other than oxygen
    3. Inorganic compound may be nitrate, sulfate or carbonate
    4. Final hydrogen acceptor is an organic compound
    Ref. 15 - p. 55

SECTION I - BACTERIAL PHYSIOLOGY

70. GENERALIZED TRANSDUCTION IS AKIN TO TRANSFORMATION:
    1. In transferring only small fragments of bacterial DNA from a donor to a recipient strain
    2. In both, the DNA is naked
    3. In both, the DNA reaches the cell as a packet surrounded by the coat of the phage
    4. The intracellular events are much the same since the coat remains outside the cell and only the DNA penetrates
    Ref. 12 - p. 205

71. WHICH OF THE FOLLOWING PROPERTIES OF NUCLEIC ACIDS ARE RELEVANT TO THE MAJOR FUNCTIONS OF GENES?:
    1. Replication
    2. Mutation
    3. Recombination
    4. Expression
    Ref. 12 - p. 210

72. BACTERIA CAN BECOME RESISTANT TO DRUGS BY THE FOLLOWING MECHANISM:
    1. Mutation
    2. Recombination
    3. Acquisition of plasmids
    4. Sensitization
    Ref. 15 - p. 47

73. DENATURATION (COLLAPSE OF THE HELICAL STRUCTURE) OF NUCLEIC ACID IS CAUSED BY:
    1. High temperature
    2. Low temperature
    3. Acid pH
    4. Substances that compete with the basis for hydrogen bond formation
    Ref. 12 - p. 215

74. THE DOUBLE-STRANDEDNESS OF DNA:
    1. Provides a mechanism for self-replication
    2. Permits damage in one strand to be corrected by removal of the damaged piece
    3. Allows messenger RNA to be formed on one strand
    4. Limits self-replication
    Ref. 12 - p. 219

75. THE FOLLOWING STRUCTURES ARE ESPECIALLY RELEVANT TO THE GENETIC ASPECT OF TRANSLATION:
    1. mRNAs
    2. Amino acid-activating enzymes
    3. tRNAs
    4. RNase
    Ref. 12 - p. 229

76. THE FOLLOWING ARE CHEMICAL MUTAGENS:
    1. Base analogs
    2. Nitrous acid
    3. Alkylating agents
    4. Hydroxylamine
    Ref. 12 - p. 242

77. SPONTANEOUS MUTATIONS ARISE IN PART FROM MUTAGENS PRESENT IN THE ENVIRONMENT, INCLUDING:
    1. Cosmic radiation
    2. Radioactive compounds
    3. Naturally occurring base analogs
    4. Chemicals
    Ref. 12 - p. 246

SECTION I - BACTERIAL PHYSIOLOGY

78. GENETIC TRANSFER IN BACTERIA CAN BE ACCOMPLISHED BY THE FOLLOWING DISTINCT MECHANISMS:
    1. Cell conjugation
    2. Viral infection
    3. Uptake of naked DNA
    4. Cell fusion
    Ref. 12 - p. 184

79. SUPPRESSOR MUTATIONS MAY ALTER:
    1. A tRNA molecule
    2. An activating enzyme
    3. The ribosomes
    4. A mRNA molecule
    Ref. 12 - p. 248

80. THE PROCESS OF INFORMATION TRANSFER FROM DNA TO PROTEIN REQUIRES THE PRESENCE OF:
    1. Ribosomes
    2. DNA
    3. tRNA
    4. Certain enzymes
    Ref. 12 - p. 272

81. THE BEST-KNOWN PLASMIDS ARE:
    1. The sex factors
    2. The Col factors
    3. The resistance (R) factors
    4. The penicillinase plasmids of staphylococci
    Ref. 15 - pp. 41-42

82. TRANSDUCTION:
    1. A large portion of donor chromosome is carried to the recipient by a temperate bacteriophage
    2. Only a small fragment of the donor's chromosome is transferred to the recipient
    3. May be generalized or restricted
    4. Occurs only in gram-negative organisms
    Ref. 15 - p. 40

83. CONCERNING TRANSFORMATION:
    1. The recipient cell takes up soluble DNA released from the donor cell
    2. Transformation does not occur in bacteria which are capable of taking up high molecular weight DNA from the medium
    3. It has been carried out only in the pneumococcus
    4. Transformable bacteria are capable of taking up DNA from any source but form genetic recombinants only if the donor is a closely related organism
    Ref. 15 - p. 40

84. IN GENERAL THE TYPES OF ALTERATIONS WHICH OCCUR IN THE NUCLEOTIDE SEQUENCE OF DNA INCLUDE:
    1. Deletions, the loss of one or more nucleotides
    2. Additions, the acquisition of one or more nucleotides
    3. Transversion, the substitution of a purine for a pyrimidine and vice versa
    4. Transitions, purine/purine or pyrimidine/pyrimidine substitutions
    Ref. 6 - p. 131

85. IN ADDITION TO CARBON AND NITROGEN, LIVING CELLS REQUIRE THE FOLLOWING FOR GROWTH:
    1. Sulfur
    2. Phosphorous
    3. Magnesium ion
    4. Ferrous ion
    Ref. 15 - p. 74

SECTION I - BACTERIAL PHYSIOLOGY

ANSWER THE FOLLOWING QUESTIONS BY USING THE KEY OUTLINED BELOW:
A. If both statement and reason are true and related cause and effect
B. If both statement and reason are true but not related cause and effect
C. If the statement is true but the reason is false
D. If the statement is false but the reason is true
E. If both statement and reason are false

86. The fimbriae in bacteria do not seem to be related to motility BECAUSE motile bacteria may be non-fimbriate, and nonmotile bacteria may possess fimbriae. Ref. 5 - p. 43

87. The term "nuclear equivalent" rather than "nucleus" is used when referring to bacteria BECAUSE no membrane has been seen between the nuclear bodies and the cytoplasm. Ref. 5 - p. 59

88. The mechanism of protein synthesis in microorganisms is better understood than in animal cells BECAUSE protein synthesis in bacteria depends on the constant production of new messenger RNA. Ref. 5 - p. 113

89. The temperate phage disappears after infecting the bacteria BECAUSE the nucleic acid of the phage is not reproduced in synchrony with the rest of the bacterial genetic material. Ref. 5 - pp. 75-76

90. Mutation is not a spontaneous occurrence BECAUSE it is not independent of the nature of the selective agent. Ref. 5 - p. 74

91. The modes of action of mutagens may differ BECAUSE they do not possess in common the ability to alter the nucleotide sequence of DNA. Ref. 6 - p. 132

92. The effects of mutation are potentially unlimited BECAUSE they may affect the properties of DNA with respect to its replicative function as well as information transfer. Ref. 6 - p. 131

93. The synthesis of messenger RNA in bacteria is not well understood BECAUSE bacteria contain several forms of RNA. Ref. 5 - p. 112

94. Most organisms grow best at a pH of 6.0 - 8.0 BECAUSE most organisms have a fairly narrow optimal pH range. Ref. 15 - p. 75

95. A gene which determines the structure of a particular protein is called a structural gene BECAUSE the activity of a structural gene is strictly regulated in the cell. Ref. 15 - p. 46

96. The bacterial cell requires a mitotic spindle BECAUSE it has a set of chromosomes to be segregated to the two daughter cells. Ref. 12 - p. 45

97. Inheritable changes in bacteria may appear very slowly BECAUSE bacteria produce enormous numbers of progeny in a short time, and their spontaneous mutants are often subject to strong selective pressures. Ref. 12 - p. 170

98. Chemical mutagenesis is important not only as a tool for studying the nature of mutations but also BECAUSE it is the closest step attained thus far toward altering the genetic material of organisms in a predictable way. Ref. 12 - p. 247

SECTION I - BACTERIAL PHYSIOLOGY

99. Most mutagenic agents do not kill microorganisms BECAUSE they do not produce changes in nucleic acid that upset its replication.
    Ref. 12 - p. 246

100. Single-stranded DNA is denser than double-stranded DNA of the same average base composition BECAUSE of the decreased hydration.
    Ref. 12 - p. 215

ANSWER THE FOLLOWING QUESTIONS BY USING THE KEY OUTLINED BELOW:
A. If only A is correct
B. If only B is correct
C. If both A and B are correct
D. If neither A nor B is correct

101. IN CONTRAST TO FUNGAL SPORES, BACTERIAL SPORES ARE:
   A. Endospores   C. Both
   B. Exospores   D. Neither
                               Ref. 12 - p. 153

102. BACTERIA THAT CAN GROW WITH OR WITHOUT AIR ARE KNOWN AS:
   A. Obligate aerobes   C. Both
   B. Obligate anaerobes   D. Neither
                               Ref. 12 - p. 65

103. PROCESS IN WHICH ENERGY IS DERIVED FROM RESPIRATION OF INORGANIC ELECTRON DONORS:
   A. Photosynthesis   C. Both
   B. Chemoautotrophy   D. Neither
                               Ref. 12 - p. 71

104. PSYCHROPHILES CAN TOLERATE TEMPERATURES UP TO:
   A. 55°
   B. 75°
   C. Both
   D. Neither   Ref. 12 - p. 135

105. THE ADDITION OF SODIUM THIOGLYCOLLATE TO A MEDIUM MARKEDLY:
   A. Decreases sensitivity to oxygen   C. Both
   B. Increases sensitivity to oxygen   D. Neither
                               Ref. 12 - p. 136

106. MEDIA ARE ENRICHED WITH MEAT EXTRACT OR YEAST EXTRACT:
   A. To provide a better source of vitamins
   B. To provide a better source of coenzymes
   C. Both
   D. Neither   Ref. 12 - p. 139

107. HEMOPHILUS IS GENERALLY GROWN ON CHOCOLATE AGAR:
   A. To release the hemin from the denatured hemoglobin
   B. To denature an enzyme that hydrolyzes DPN
   C. Both
   D. Neither   Ref. 12 - p. 139

108. CELL MASS CAN BE DETERMINED BY:
   A. Dry weight   C. Both
   B. Turbidity   D. Neither
                               Ref. 12 - p. 141

SECTION I - BACTERIAL PHYSIOLOGY

109. IN THE APPROACH TO AND DURING THE STATIONARY PHASE, BACTERIAL CELLS BECOME:
    A. Larger
    B. Smaller
    C. Both
    D. Neither
    Ref. 12 - p. 142

110. WHEN MASS INCREASES LINEARLY WITH TIME, THE GROWTH IS CALLED:
    A. Lag phase
    B. Stationary phase
    C. Both
    D. Neither
    Ref. 12 - p. 143

111. IN THE PNEUMOCOCCUS, THE R TO S TRANSITION IS ASSOCIATED WITH THE ACQUISITION OF:
    A. A capsule
    B. A wall polysaccharide
    C. Both
    D. Neither
    Ref. 12 - p. 149

112. THE UNIQUE FEATURE OF SPORES IS:
    A. Dehydration
    B. Huge content of Ca^{++}
    C. Both
    D. Neither
    Ref. 12 - p. 155

113. TRANSFORMATION INVOLVES:
    A. Replacement
    B. Recombination
    C. Both
    D. Neither
    Ref. 12 - p. 187

114. THE GENETIC INFORMATION OF SOME VIRUSES IS STORED IN:
    A. Single-stranded DNA
    B. Single-stranded RNA
    C. Both
    D. Neither
    Ref. 12 - p. 210

115. A SPECIFIC FORM OF A GENE IS CALLED AN ALLELE WHICH CAN BE OF:
    A. Wild-type
    B. Mutant
    C. Both
    D. Neither
    Ref. 12 - p. 219

116. X-IRRADIATION CAUSES:
    A. Excitations of chemical groups in DNA
    B. Production of highly reactive and short-lived radicals in the water surrounding the DNA
    C. Both
    D. Neither
    Ref. 12 - p. 258

117. THE RATE OF PROTEIN SYNTHESIS IN BALANCED GROWTH IS PROPORTIONAL TO THE:
    A. Number of ribosomes present
    B. Amount of transfer components
    C. Both
    D. Neither
    Ref. 12 - p. 289

118. WHEN BACTERIAL CELLS ARE TRANSFERRED FROM GROWTH MEDIUM TO A MEDIUM LACKING A REQUIRED AMINO ACID OR PYRIMIDINE, THE:
    A. DNA increases
    B. Cell number increases
    C. Both
    D. Neither
    Ref. 12 - p. 294

SECTION I - BACTERIAL PHYSIOLOGY

119. SOME PROTEIN SYNTHESIS IS REQUIRED FOR:
    A. The initiation of each new round of DNA replication
    B. The completion of a round
    C. Both
    D. Neither
    Ref. 12 - p. 293

120. A PHENOTYPIC ADAPTATION TO AN ENVIRONMENTAL CHANGE:
    A. Involves essentially all the cells in culture
    B. Involves a rare mutant which is then selected
    C. Both
    D. Neither
    Ref. 12 - p. 170

FOR EACH OF THE FOLLOWING MULTIPLE CHOICE QUESTIONS CHOOSE THE ONE MOST APPROPRIATE ANSWER:

121. LABORATORY WORK WITH BACTERIA:
    A. Does not necessitate sterilization of equipment
    B. Filtration is the best method for sterilization
    C. Should be performed with pure cultures
    D. Does not require isolation
    E. Depends only on biochemical differences
    Ref. 2 - p. 13

122. GLASSWARE AND INSTRUMENTS ARE STERILIZED BY:
    A. Dipping in iodine solutions
    B. Dipping in alcohol
    C. Thorough cleansing
    D. Hot air oven
    E. Rinsing with acid
    Ref. 2 - p. 13

123. PLATINUM WIRE NEEDLES ARE MOST CONVENIENTLY STERILIZED BY:
    A. Heating to a dull red in a Bunsen burner flame
    B. Autoclaving
    C. Dry heat
    D. Intermittent sterilization
    E. Alcohol
    Ref. 2 - p. 13

124. BOILING:
    A. Kills bacterial spores
    B. Is adequate for sterilizing surgical instruments
    C. Is inadequate for destroying the spores of bacteria and fungi
    D. Is as effective as intermittent sterilization
    E. Is the method of choice for sterilizing syringes and needles
    Ref. 2 - p. 13

125. INTERMITTENT STERILIZATION IS THE METHOD OF CHOICE:
    A. For all laboratory equipment
    B. For all laboratory media
    C. For certain glass equipment
    D. Because it takes only 30 minutes
    E. None of the above
    Ref. 2 - p. 13

126. THE AUTOCLAVE METHOD OF STERILIZATION:
    A. Raises the temperature to 120° C.
    B. Is performed under atmospheric pressure
    C. Differs in theory from a pressure cooker
    D. Utilizes dry heat
    E. None of the above
    Ref. 2 - p. 14

SECTION I - BACTERIAL PHYSIOLOGY

127. STERILIZATION BY FILTRATION:
   A. Is not particularly useful for toxins
   B. Is not accomplished by differential pressure
   C. Is recommended for viruses
   D. Is useful for soluble products such as toxin
   E. Is useful for rickettsiae   Ref. 2 - p. 14

128. THE BASAL NUTRIENT MEDIUM FOR MICROORGANISMS:
   A. Always contains agar
   B. Does not contain any inhibitory substance
   C. May be enriched to be both selective and inhibitory
   D. Is always solid
   E. Is always liquid   Ref. 2 - p. 14

129. NITRATE BROTH IS USED TO TEST:
   A. The ability of the bacteria to reduce nitrate
   B. For nitrogen fixation
   C. For insoluble carbohydrate
   D. For the presence of indole
   E. For the presence of H_2S Ref. 2 - p. 15

130. TRYPTOPHAN BROTH IS USED TO:
   A. Enhance growth of E. coli
   B. Delay growth of Proteus vulgaris
   C. Stop overgrowth of Staphylococcus aureus
   D. Test indole formation from tryptophan
   E. Test for formation of acid from sugar
       Ref. 2 - p. 15

131. LEAD ACETATE AGAR:
   A. Is inoculated by spreading the unknown material on the surface
   B. Is used to test for formation of lead
   C. Turns blue if hydrogen sulfide is produced
   D. Turns brown or black if hydrogen sulfide is produced
   E. Turns red if hydrogen peroxide is formed
       Ref. 2 - p. 15

132. BLOOD AGAR:
   A. Tends to inhibit growth since blood is bacteriostatic
   B. Is prepared by mixing blood and agar 1:1
   C. Is rarely used in diagnostic work
   D. Is used to test for hemolysis
   E. Sheep blood cannot be used   Ref. 2 - p. 16

133. CYSTINE BLOOD AGAR IS USEFUL PRIMARILY IN THE CULTIVATION OF:
   A. Pasteurella tularensis   D. Streptococcus
   B. Pneumococcus   E. None of the above
   C. Pasteurella pestis   Ref. 2 - p. 16

134. CHOCOLATE AGAR:
   A. Is made with a derivative of cocoa
   B. Is prepared by boiling blood
   C. Is particularly useful for the cultivation of gonococcus and meningococcus
   D. Is useful in culturing C. diphtheriae
   E. Is used to demonstrate hemolysis   Ref. 2 - p. 16

SECTION I - BACTERIAL PHYSIOLOGY

135. LÖFFLER'S MEDIUM IS USED PRIMARILY TO CULTURE:
    A. Tubercle bacillus
    B. Gonococcus
    C. Brucella abortus
    D. Diphtheria bacillus
    E. None of the above
    Ref. 2 - p. 16

136. PETRAGNANI'S MEDIUM:
    A. Contains gentian violet
    B. Has no inhibiting substance
    C. Is used to culture tubercle bacillus
    D. Is used for tissue culture
    E. Precipitates lead acetate
    Ref. 2 - p. 18

137. MOTILITY:
    A. Is not a characteristic of differential value
    B. Can be detected by gross examination of the culture on all media
    C. Is demonstrated by the hanging drop method
    D. Is a feature of all staphylococci
    E. Can be positively identified by staining methods
    Ref. 2 - p. 18

138. THE GRAM STAIN:
    A. Is not really a differential stain
    B. Consists of crystal violet, iodine, alcohol and counterstain in that order
    C. Consists of crystal violet, alcohol, iodine and counterstain in that order
    D. Consists of counterstain, iodine, alcohol, and crystal violet in that order
    E. Is red when positive, blue when negative
    Ref. 2 - p. 20

139. ACID-FAST BACTERIA:
    A. Are characterized by a low lipid content
    B. Are so called because acid dyes are quicker
    C. Are stained with the Ziehl-Neelsen method
    D. Are represented by the corynebacteria
    E. Are stained best by simple stain
    Ref. 2 - p. 20

140. HISS' METHOD:
    A. Stains flagella
    B. Stains nuclei
    C. Stains acid-fast bacteria
    D. Stains capsules
    E. None of the above
    Ref. 2 - pp. 20-21

141. INCREASED CARBON DIOXIDE TENSION:
    A. Is needed to grow P. pestis
    B. Is not needed by any bacteria
    C. May be obtained by placing a short candle stub and the culture in a jar, lighting it, and closing the jar
    D. Results in a chlorophyll shift
    E. Is incorporated, when needed, into the medium
    Ref. 2 - p. 22

142. ROUTINE METHODS USED FOR THE CULTIVATION OF ANAEROBIC BACTERIA INCLUDE:
    A. The use of boiled medium in a deep tube
    B. Chemical absorption of oxygen
    C. Removal of oxygen by combustion
    D. Displacement of air by an inert gas
    E. All of the above
    Ref. 2 - pp. 22-23

SECTION I - BACTERIAL PHYSIOLOGY

143. THE FOLLOWING METHOD IS USED AS A PRESERVATIVE MEASURE THEREBY PREVENTING PENETRATION OF FOOD BY BACTERIA:
    A. Heat
    B. Salt
    C. Sugar
    D. Acids
    E. All of the above
    Ref. 15 - p. 91

144. PASTEURIZATION MAY BE CARRIED OUT BY MAINTAINING THE MILK AT:
    A. 50° C for 30 minutes
    B. 62° C for 30 minutes
    C. 100° C for 30 minutes
    D. 120° C for 30 minutes
    E. None of the above
    Ref. 15 - p. 89

145. ZERO GROWTH RATE IS OBSERVED DURING THE FOLLOWING PHASE:
    A. Lag
    B. Acceleration
    C. Exponential
    D. Retardation
    E. Decline
    Ref. 15 - p. 80

146. THE MOST RELIABLE METHOD OF STERILIZATION:
    A. Freezing
    B. Heat
    C. X-irradiation
    D. Sonic vibrations
    E. Filtration
    Ref. 6 - p. 212

147. THE MOST EFFECTIVE METHOD OF HEAT DISINFECTION KNOWN:
    A. Hot air
    B. Boiling
    C. Live steam
    D. Steam under pressure
    E. Pasteurization
    Ref. 6 - p. 213

148. DRY HEAT STERILIZATION IS COMMONLY USED FOR:
    A. Oils
    B. Powders
    C. Glassware
    D. Jellies
    E. All of the above
    Ref. 6 - p. 213

149. ULTRAVIOLET RADIATION:
    A. The most effective bactericidal wavelength is 240 to 280 nm
    B. Leads to inhibition of DNA synthesis
    C. The energy is low
    D. The power of penetration is very poor
    E. All of the above
    Ref. 6 - p. 215

150. THE FOLLOWING FILTER IS MADE OF DIATOMACEOUS EARTH:
    A. Seitz
    B. Berkefeld
    C. Selas
    D. Chamberland
    E. None of the above
    Ref. 6 - p. 217

ANSWER THE FOLLOWING QUESTIONS BY USING THE KEY OUTLINED BELOW:
A. If only A is correct
B. If only B is correct
C. If both A and B are correct
D. If neither A nor B is correct

151. THE PHASE MICROSCOPE HAS BEEN PARTICULARLY USEFUL IN STUDYING THE STRUCTURES OF:
    A. Living bacteria
    B. Stained bacteria
    C. Both
    D. Neither
    Ref. 2 - p. 38

SECTION I - BACTERIAL PHYSIOLOGY 25

152. THE DARKFIELD MICROSCOPE HAS CONSIDERABLE UTILITY IN THE STUDY OF:
    A. Treponema
    B. Leptospira
    C. Both
    D. Neither
    Ref. 2 - p. 38

153. THE COMMON ROUTES OF INOCULATION OF EXPERIMENTAL ANIMALS FOR THE STUDY OF PATHOGENIC BACTERIA INCLUDE:
    A. Intradermal
    B. Intraperitoneal
    C. Both
    D. Neither
    Ref. 2 - p. 24

154. IN THE GRAM STAIN, THE FOLLOWING PROCEDURE IS HIGHLY SPECIFIC:
    A. Alcohol
    B. Counterstain
    C. Both
    D. Neither
    Ref. 2 - p. 20

155. USED IN TISSUE CULTURES TO PREVENT MICROBIAL CONTAMINATION:
    A. Penicillin G
    B. Streptomycin
    C. Both
    D. Neither
    Ref. 2 - p. 28

156. THE MASSES OF DNA IN BACTERIA STAIN STRONGLY WITH:
    A. Giemsa stain
    B. Feulgen stain
    C. Both
    D. Neither
    Ref. 2 - p. 57

157. PIGMENTATION IS COMMON AMONG:
    A. Saprophytic bacteria
    B. Pathogenic bacteria
    C. Both
    D. Neither
    Ref. 2 - p. 63

158. THE ABILITY OF THE BACTERIA TO REDUCE TELLURIUM SALTS WHICH GIVE RISE TO BLACK COLONIES ON TELLURITE DIFFERENTIAL MEDIUM:
    A. Tubercle bacilli
    B. Diptheria bacilli
    C. Both
    D. Neither
    Ref. 2 - p. 64

159. HAIR-LIKE FILAMENTS EXTENDING FROM THE CELLS ARE KNOWN AS:
    A. Pili
    B. Fimbriae
    C. Both
    D. Neither
    Ref. 6 - p. 32

160. BACTERIA WITH A SINGLE FLAGELLUM AT ONE END ARE CALLED:
    A. Amphitrichous
    B. Lophotrichous
    C. Both
    D. Neither
    Ref. 6 - p. 30

161. ENDOSPORE FORMATION IS A DISTINGUISHING FEATURE OF ORGANISMS OF THE GENUS:
    A. <u>Bacillus</u>
    B. <u>Clostridium</u>
    C. Both
    D. Neither
    Ref. 6 - p. 44

SECTION I - BACTERIAL PHYSIOLOGY

162. GROWTH OF BACTERIA INVOLVES:
   A. Growth of the individual bacterium
   B. Growth of populations or cultures
   C. Both
   D. Neither
   Ref. 6 - pp. 69-70

163. ABSENCE OF A CELL WALL IS CHARACTERISTIC OF:
   A. Spirochetes
   B. Mycoplasmas
   C. Both
   D. Neither
   Ref. 15 - p. 25

164. THE WALLS OF GRAM-POSITIVE BACTERIA CONSIST OF:
   A. Lipopolysaccharide
   B. Lipoprotein
   C. Both
   D. Neither
   Ref. 15 - p. 11

165. THE AXIAL FILAMENT OF THE SPIROCHETE:
   A. Lies between the cell membrane and cell wall
   B. Its contraction may be responsible for motility
   C. Both
   D. Neither
   Ref. 15 - p. 31

166. THE BACTERIAL ENDOSPORES:
   A. Are metabolically inert
   B. Do not contain any enzymes
   C. Both
   D. Neither
   Ref. 15 - pp. 14-15

167. IN GENERAL, BACTERIA REPRODUCE BY:
   A. Binary fission
   B. Budding
   C. Both
   D. Neither
   Ref. 15 - p. 21

168. THE ROUTES OF INOCULATION OF FERTILE EGGS FOR THE PROPAGATION OF VIRUSES:
   A. Amniotic cavity
   B. Chorioallantoic membrane
   C. Both
   D. Neither
   Ref. 2 - pp. 26-27

169. BROAD-SPECTRUM ANTIBIOTICS MUST NOT BE GIVEN TO FLOCKS IF THE EGGS ARE USED FOR THE PROPAGATION OF:
   A. Rickettsiae
   B. Psittacosis-lymphogranuloma venereum group
   C. Both
   D. Neither
   Ref. 2 - p. 26

170. MOST COMMONLY USED CONTINUOUS CELL LINE CULTURE FOR VIRAL STUDIES:
   A. Monkey kidney tissue culture
   B. HeLa cell culture
   C. Both
   D. Neither
   Ref. 2 - p. 28

FOR EACH OF THE FOLLOWING MULTIPLE CHOICE QUESTIONS, CHOOSE THE ONE MOST APPROPRIATE ANSWER:

171. ANTISEPTICS AND DISINFECTANTS:
   A. Are the same
   B. Destroy all bacteria
   C. Are both used for living tissue
   D. Are both highly toxic to living tissue
   E. Antiseptics are non-toxic to living tissue, disinfectants may be toxic
   Ref. 2 - p. 185

SECTION I - BACTERIAL PHYSIOLOGY

172. STRONG ACIDS AND ALKALIS:
     A. Have a bacteriostatic effect
     B. Have a bactericidal effect
     C. Have no relationship to the activity of other disinfecting agents
     D. Are both contained in lysol
     E. None of the above   Ref. 2 - p. 185

173. IN GENERAL, SALTS:
     A. Are all bacteriostatic
     B. Are all bactericidal
     C. May stimulate growth in low concentration
     D. Of lighter metals are equally toxic as those of heavier metals
     E. Have no effect on bacterial growth   Ref. 2 - p. 185

174. OXIDIZING AGENTS IN THE FORM OF CHLORAMINES:
     A. Do not show disinfectant properties
     B. Show weak disinfectant properties
     C. Are only bacteriostatic
     D. Are represented by chloramine-T and dichloramine-T
     E. Have not been of value in the disinfection of deep wounds
                                         Ref. 2 - pp. 186-187

175. THE FOLLOWING IS NOT GENERALLY USED AS A DISINFECTANT:
     A. Phenol
     B. Aldehyde
     C. Soap
     D. Quaternary ammonium compound
     E. Alcohol   Ref. 2 - p. 187

176. THE PHENOL COEFFICIENT:
     A. Refers to the solubility of phenol in water
     B. Refers to the anesthetic properties of phenol
     C. Is a method of comparing various bacteria
     D. Is a standardized method of comparing the efficiency of a given
        disinfectant to that of phenol
     E. Is stable for a disinfectant despite the type of bacteria
                                         Ref. 2 - p. 195

177. HEXACHLOROPHENE IS:
     A. A phenol derivative
     B. A metal-organic compound
     C. A detergent
     D. An ineffective commercial product
     E. None of the above   Ref. 2 - p. 189

178. THE FOLLOWING ALKYLATING AGENT IS A HIGHLY EFFECTIVE
     BACTERICIDE:
     A. Ethyl alcohol   D. Carbon monoxide
     B. Diethyl ether   E. Methyl alcohol
     C. Ethylene oxide   Ref. 2 - p. 190

179. THE FOLLOWING DYE IS BACTERIOSTATIC:
     A. Malachite green   D. All of the above
     B. Brilliant green   E. None of the above
     C. Basic fuchsin   Ref. 2 - p. 191

SECTION I - BACTERIAL PHYSIOLOGY

180. THE ACTION OF VARIOUS DISINFECTANTS ON BACTERIA:
    A. Is non-specific
    B. Is about the same since they are of equal strength
    C. Is not different from their action on viruses
    D. Varies with the particular disinfectant and bacteria
    E. Is purely a chemical reaction   Ref. 2 - p. 193

181. THE LEAST IMPORTANT FACTOR INFLUENCING DISINFECTION IS THE:
    A. Concentration of the disinfectant
    B. Presence of extraneous organic matter
    C. Temperature
    D. Presence of salts
    E. pH   Ref. 2 - p. 193

182. THE SULFONES ARE MOST USEFUL IN THE TREATMENT OF:
    A. Tuberculosis
    B. Leprosy
    C. Gonorrhea
    D. Meningitis
    E. Streptococcal infections
    Ref. 2 - p. 198

183. THIOSEMICARBOZONE COMPOUNDS ARE ACTIVE AGAINST THE:
    A. Tubercle bacillus
    B. Rickettsiae
    C. Polioviruses
    D. Salmonella group
    E. Treponema group
    Ref. 2 - p. 199

184. THE SULFONAMIDE COMPOUNDS ARE RELATIVELY INEFFECTIVE AGAINST:
    A. The plague bacillus
    B. The dysentery bacillus
    C. The typhoid bacillus
    D. Streptococci
    E. Gonococcus
    Ref. 2 - p. 201

185. THE MOST PROLIFIC SOURCE OF CHEMOTHERAPEUTIC ANTIBIOTICS HAS BEEN THE:
    A. Bacteria
    B. Sporulating bacteria
    C. Licheniformin
    D. Higher fungi
    E. Rickettsiae
    Ref. 2 - p. 205

186. ANTIBIOTIC SUBSTANCES FROM BACTERIA DO NOT INCLUDE:
    A. Subtilin
    B. Streptomyces
    C. Licheniformin
    D. Colicin
    E. Bacitracin
    Ref. 2 - p. 205

187. PENICILLIN IS:
    A. A single substance
    B. Activated by penicillinase
    C. Formed by many species of *Penicillium*
    D. Most active against gram-negative bacteria
    E. Relatively stable   Ref. 2 - pp. 206-207

188. STREPTOMYCIN:
    A. Toxicity is not manifested as an effect on the eighth nerve
    B. Is ineffective on the tubercle bacillus
    C. Does not have the disadvantage of having bacteria become resistant to it easily
    D. Is effective in the treatment of staphylococcal infections
    E. Is effective against gram-negative bacteria
       Ref. 2 - p. 207

SECTION I - BACTERIAL PHYSIOLOGY

189. THE GREATEST DISADVANTAGE OF CHLORAMPHENICOL IS:
    A. The production of severe nausea and vomiting
    B. Slow and incomplete absorption from the gastrointestinal tract
    C. Its ineffectiveness when given by mouth
    D. Development of aplastic anemia following the administration of this drug
    E. Its high toxicity
    Ref. 2 - p. 208

190. THE ISOMORPHIC TETRACYCLINE COMPOUNDS ARE:
    A. Chloramphenicol and aureomycin
    B. Chloromycetin and terramycin
    C. Aureomycin and terramycin
    D. Penicillin and aureomycin
    E. None of the above
    Ref. 2 - p. 208

191. TETRACYCLINES INCLUDE ALL OF THE FOLLOWING, EXCEPT:
    A. Chloramphenicol
    B. Chlortetracycline
    C. Oxytetracycline
    D. Terramycin
    E. None of the above
    Ref. 2 - p. 208

192. THE ANTIBIOTIC OF CHOICE FOR TYPHOID FEVER IS:
    A. Chloramphenicol
    B. Sulfadiazine
    C. Penicillin
    D. Streptomycin
    E. Neomycin
    Ref. 2 - p. 207

193. SYNERGISTIC EFFECT OF ANTIBACTERIAL DRUG COMBINATIONS OCCURS IF THE ACTIVITY OF:
    A. Two drugs in combination is greater than that obtained by doubling the concentration of either alone
    B. The combination is greater than that of either drug alone, but less than that obtained by doubling the concentration of either
    C. The combined drugs is not greater than that of either drug alone
    D. The combination is less than that obtained with either drug alone
    E. None of the above
    Ref. 2 - p. 214

194. DISADVANTAGES OF CHLORTETRACYCLINE ADMINISTRATION:
    A. Highly toxic
    B. Causes severe nausea
    C. Causes severe vomiting
    D. Causes severe diarrhea
    E. None of the above
    Ref. 2 - p. 208

195. ISONIAZID HAS HAD A DRAMATIC EFFECT ON:
    A. Leprosy
    B. Typhoid fever
    C. Staphylococcal infections
    D. Tuberculosis
    E. None of the above
    Ref. 12 - p. 324

196. THE MECHANISM OF ACTION OF PENICILLIN:
    A. Interference with membrane function
    B. Interference with cell-wall synthesis
    C. Interference with protein synthesis
    D. Interference with nucleic acid metabolism
    E. Interference with intermediary metabolism
    Ref. 6 - p. 176

197. ANTIBIOTICS HAVE A WIDE DISTRIBUTION IN NATURE WHERE THEY PLAY AN IMPORTANT ROLE IN REGULATING THE MICROBIAL POPULATION OF:
    A. Soil
    B. Water
    C. Sewage
    D. All of the above
    E. None of the above
    Ref. 6 - p. 173

SECTION I - BACTERIAL PHYSIOLOGY

198. AS A HAND WASH ETHYL ALCOHOL IS RECOMMENDED IN THE:
  A. 60% solution
  B. 70% solution
  C. 80% solution
  D. 90% solution
  E. 100% solution
  Ref. 3 - p. 276

199. $CuSO_4$ IS USED TO:
  A. Preserve serum and vaccines
  B. Disinfect instruments
  C. Prevent newborn ophthalmia
  D. Suppress algae in water reservoirs
  E. Purify drinking water
  Ref. 3 - p. 275

200. WHICH ONE OF THE FOLLOWING STATEMENTS IS NOT TRUE CONCERNING BACITRACIN?:
  A. It is active against gram-positive bacteria
  B. It is inactive against gram-negative bacteria
  C. It destroys susceptible bacteria by attacking the cell membrane
  D. The source organism is B. subtilis
  E. It is quite toxic if given internally
  Ref. 3 - p. 301

201. THE CONDITION MOST FAVORABLE FOR GROWTH OF PENICILLIUM NOTATUM AND PENICILLIN PRODUCTION:
  A. Anaerobic
  B. 40° C
  C. pH 5.5-6.0
  D. Aerobic
  E. Presence of chrysogenin
  Ref. 3 - p. 289

202. PENICILLIN IS:
  A. Activated by alcohol
  B. Unstable in the crystalline form
  C. Unaffected by the acidity of gastric juice
  D. Not readily soluble in water
  E. A strong monobasic acid
  Ref. 3 - p. 290

203. PENICILLIN SUSCEPTIBLE BACTERIA INCLUDE ALL OF THE FOLLOWING, EXCEPT:
  A. Streptococcus
  B. Pneumococcus
  C. Meningococcus
  D. Salmonella typhosa
  E. Treponema pallidum
  Ref. 3 - p. 294

204. THE FOLLOWING METHOD FOR ASSAYING THE POTENCY OF, OR SENSITIVITY OF MICROORGANISMS TO ANTIBIOTICS IS PROBABLY NOT VERY EXACTING BUT IS PREFERRED AND WIDELY USED FOR EXPEDIENCY AND ECONOMICS:
  A. Serial dilution method in broth
  B. Serial dilution method in agar
  C. Cylinder-plate method
  D. Disk or tablet method
  E. Growth-indicator method
  Ref. 3 - p. 296

205. WHICH OF THE FOLLOWING STATEMENTS REFERRING TO PENICILLIN RESISTANCE IS INCORRECT?:
  A. The major cause of resistance to penicillin is penicillinase which inactivates the drug
  B. Penicillinase converts the antibiotic to the inactive penicilloic acid
  C. Penicillinase is rapidly produced in Staphylococcus aureus
  D. Penicillinase is also produced by enteric bacteria
  E. Penicillin is not attacked by any other enzyme
  Ref. 6 - p. 177

SECTION I - BACTERIAL PHYSIOLOGY

206. THE FOLLOWING METHOD IS USED FOR PRELIMINARY ASSAYS OF CRUDE MATERIAL IN THE SEARCH FOR NEW ANTIBIOTICS BECAUSE IT IS RAPID AND CONVENIENT:
   A. Serial dilution method in broth
   B. Precipitation method
   C. Serial dilution method in agar
   D. Cylinder plate method
   E. Blood agar plate method
   Ref. 3 - p. 295

207. ALL OF THE FOLLOWING ANTIBIOTICS ARE DERIVED FROM SPECIES OF STREPTOMYCES, EXCEPT:
   A. Erythromycin
   B. Oxytetracycline
   C. Bacitracin
   D. Nystatin
   E. Chloramphenicol
   Ref. 3 - p. 299

208. STREPTOMYCIN:
   A. Numerous salts have been prepared but are of little therapeutic value
   B. It is more stable than penicillin, both in dry form and in solution
   C. It is more stable than penicillin in only the dry form
   D. Boiling does not produce loss of potency
   E. It retains its potency in the refrigerator indefinitely
   Ref. 3 - p. 298

209. THE MECHANISM OF ACTION OF THE FOLLOWING ANTIMICROBIAL DRUG IS THE INHIBITION OF CELL MEMBRANE FUNCTION:
   A. Penicillins
   B. Polymyxins
   C. Chloramphenicol
   D. Sulfonamides
   E. None of the above
   Ref. 15 - p. 107

210. PENICILLIN AND STREPTOMYCIN DIFFER FROM EACH OTHER IN ALL THE FOLLOWING RESPECTS, EXCEPT:
   A. Bacterial spectrum
   B. Stability
   C. Rate of excretion from the body
   D. Source
   E. Bactericidal action in vitro
   Ref. 3 - p. 298

211. THE ERYTHROMYCINS:
   A. Are obtained from Streptomyces venezuelae
   B. Inhibit gram-positive organisms
   C. Do not cause undesirable side-effects
   D. Resistant mutants do not emerge
   E. Are more effective than penicillin in most penicillin-sensitive infections
   Ref. 15 - p. 121

212. AMPHOTERICIN B:
   A. Is not produced by a Streptomyces species
   B. Appears to be the most effective agent for some of the systemic fungi
   C. Does not produce toxic effects
   D. All of the above
   E. None of the above
   Ref. 15 - p. 121

213. BACITRACIN IS A POLYPEPTIDE OBTAINED FROM A STRAIN OF:
   A. Bacillus polymyxa
   B. Bacillus subtilis
   C. Bacillus brevis
   D. Bacillus cereus
   E. Bacillus circulans
   Ref. 15 - p. 121

214. ALL OF THE FOLLOWING STATEMENTS REFER TO BACITRACIN, EXCEPT:
   A. It is mainly active against gram-negative bacteria
   B. Its best use is for topical application
   C. It has no place in systemic therapy
   D. It is toxic for the kidney
   E. It is mainly bactericidal for gram-positive bacteria
   Ref. 15 - p.

SECTION I - BACTERIAL PHYSIOLOGY

215. AN ANTIBIOTIC WHICH IS EFFECTIVE AGAINST PATHOGENIC FUNGI IS:
    A. Neomycin
    B. Viomycin
    C. Nystatin
    D. Carbomycin
    E. Polymyxin B
    Ref. 3 - p. 299

216. WHICH OF THE FOLLOWING STATEMENTS IS INCORRECT FOR VANCOMYCIN?:
    A. It is an antibiotic of low molecular weight
    B. It is derived from Streptomyces orientalis
    C. It is poorly absorbed from the intestine
    D. Its primary use has been in the treatment of infections due to drug resistant staphylococci
    E. Drug resistant strains do not emerge rapidly
    Ref. 15 - p. 121

217. THE CENTRAL MOLECULE OF THE SULFONAMIDE DRUGS IS:
    A. Para-amino benzoic acid
    B. Nicotinamide
    C. Isoniazid
    D. Variable
    E. Sulfanilamide
    Ref. 3 - p. 283

218. A NUMBER OF ANTIBIOTICS HAVE BEEN USED IN COMMERCIAL STOCK FEEDS TO:
    A. Enhance the growth rate of certain animals
    B. Encourage development of drug-dependent bacteria
    C. Keep animals germ-free
    D. Enhance the quality of the animal offspring
    E. Eliminate tetanus spores from animal stools
    Ref. 3 - p. 302

219. ISONIAZID:
    A. Is most active against mycobacteria, especially M. tuberculosis
    B. Is best when employed in combination with other antimycobacterial agents
    C. Is rapidly and completely absorbed from the gastrointestinal tract
    D. Freely diffuses into tissue fluids
    E. All of the above
    Ref. 15 - p. 124

220. THE TETRACYCLINES:
    A. Are effective only against gram-positive organisms
    B. Are derived from chloramphenicol
    C. Have dissimilar molecular structures
    D. Are called "broad spectrum" antibiotics
    E. Have dissimilar therapeutic range   Ref. 3 - p. 300

ANSWER THE FOLLOWING QUESTIONS BY USING THE KEY OUTLINED BELOW:
A. If only A is correct
B. If only B is correct
C. If both A and B are correct
D. If neither A nor B is correct

221. ANTIBIOTICS WHICH KILL BACTERIA ARE:
    A. Bacteriostatic
    B. Bactericidal
    C. Both
    D. Neither
    Ref. 6 - p. 173

SECTION I - BACTERIAL PHYSIOLOGY

222. THE RATE OF CHEMICAL DISINFECTION IS INFLUENCED BY:
    A. Temperature
    B. Concentration of disinfectant
    C. Both
    D. Neither
    Ref. 6 - pp. 205-206

223. THE MECHANISM BY WHICH DRUGS KILL OR INHIBIT THE GROWTH OF MICROORGANISM:
    A. Damage to the cell membrane
    B. Induction of extensive breakdown of RNA
    C. Both
    D. Neither
    Ref. 6 - p. 207

224. THE SULFONAMIDES INTERFERE WITH THE:
    A. Synthesis of the essential metabolite, p-aminobenzoic acid
    B. Cytochrome system
    C. Both
    D. Neither
    Ref. 6 - p. 197

225. GRISEOFULVIN IS A FUNGISTATIC AGENT:
    A. Specific for fungi whose walls contain chitin
    B. Used successfully against infections caused by dermatophytes
    C. Both
    D. Neither
    Ref. 6 - p. 181

226. MECHANISM OF ACTION OF PENICILLIN:
    A. Lyses growing cells
    B. Has no effect on resting cells
    C. Both
    D. Neither
    Ref. 12 - p. 315

227. PENICILLINASE, AN ENZYME THAT SPLITS THE LACTAM BOND OF PENICILLIN, IS PRODUCED BY:
    A. E. coli
    B. Staphylococcus
    C. Both
    D. Neither
    Ref. 12 - p. 316

228. CYCLOSERINE (OXAMYCIN):
    A. Is produced by most bacteria
    B. Has not been synthesized
    C. Both
    D. Neither
    Ref. 12 - p. 318

229. NOVOBIOCIN:
    A. Is bacteriostatic against most sensitive organisms
    B. Inhibits DNA and RNA synthesis in cells
    C. Both
    D. Neither
    Ref. 12 - p. 319

230. THE CONCENTRATIONS OF STREPTOMYCIN REQUIRED FOR KILLING SENSITIVE BACTERIA INCREASE WITH:
    A. Increasing acidity
    B. Increasing concentrations of salts
    C. Both
    D. Neither
    Ref. 12 - p. 322

231. INHIBITION OF PROTEIN SYNTHESIS ON THE RIBOSOME IS RESPONSIBLE FOR THE ACTION OF:
    A. Chloramphenicol
    B. Tetracyclines
    C. Both
    D. Neither
    Ref. 12 - p. 325

SECTION I - BACTERIAL PHYSIOLOGY

232. ACTINOMYCIN D:
     A. Blocks the synthesis of RNA on a DNA template
     B. Blocks DNA replication
     C. Both
     D. Neither   Ref. 12 - p. 328

233. FORMALDEHYDE IS A GAS THAT IS MARKETED AS A:
     A. 10 per cent aqueous solution   C. Both
     B. 100 per cent aqueous solution   D. Neither
                                          Ref. 12 - p. 347

234. THE MOST EFFECTIVE CATIONIC DETERGENTS ARE THE QUARTERNARY COMPOUNDS WHICH ARE USED FOR:
     A. Skin antisepsis   C. Both
     B. Sanitizing food utensils   D. Neither
                                   Ref. 12 - p. 348

235. ORGANIC MERCURY COMPOUNDS ARE USED AS:
     A. Highly irritating antiseptics
     B. Preservatives for sera and vaccines
     C. Both
     D. Neither   Ref. 12 - p. 346

236. THE BACTERICIDAL ACTION OF PHENOL DEPENDS ON:
     A. Membrane damage   C. Both
     B. Cell lysis   D. Neither
                                   Ref. 12 - p. 349

237. MECHANISM OF ACTION OF DETERGENTS:
     A. Dissolves lipids   C. Both
     B. Denatures proteins in solution   D. Neither
                                          Ref. 12 - p. 349

238. ETHANOL HAS THE ADVANTAGE OVER ISOPROPYL ALCOHOL OF BEING:
     A. Less volatile   C. Both
     B. Slightly more potent   D. Neither
                                   Ref. 12 - p. 350

239. HEXACHLOROPHENE:
     A. Is bacteriostatic in very high dilutions
     B. Is widely used as a skin antiseptic
     C. Both
     D. Neither   Ref. 12 - p. 350

240. ETHANOL IS MOST EFFECTIVE IN:
     A. 95 per cent aqueous solution
     B. 100 per cent
     C. Both
     D. Neither   Ref. 12 - p. 350

SECTION II - IMMUNOLOGY

FOR EACH OF THE FOLLOWING MULTIPLE CHOICE QUESTIONS, CHOOSE THE ONE MOST APPROPRIATE ANSWER:

241. THE PREDOMINANT ANTIBODY OF A SECONDARY IMMUNE RESPONSE IS:
    A. γG
    B. γM
    C. γA
    D. γE
    E. γD
    Ref. 16 - p. 53

242. THE FOLLOWING TYPE OF CELL IS BELIEVED TO HAVE TWO FUNCTIONS - ANTIBODY PRODUCTION AND CELL-MEDIATED INTERACTIONS:
    A. Monocytes
    B. Neutrophils
    C. Eosinophils
    D. Lymphocytes
    E. Basophils
    Ref. 16 - p. 24

243. ANAMNESTIC REACTION REFERS TO:
    A. Gradually rising antibody titers
    B. True immunologic paralysis
    C. The prompt production of antibodies following secondary injections of antigen
    D. The lag period in antibody production following a single primary injection of antigen
    E. Species specific antibodies
    Ref. 10 - p. 93

244. THE REACTION THAT OCCURS WHEN ANTIBODY AND SOLUBLE ANTIGEN ARE MIXED IS DEMONSTRATED BY THE:
    A. Agglutination test
    B. Precipitin test
    C. Adsorption test
    D. Complement-fixation test
    E. Hemagglutination test
    Ref. 10 - p. 349

245. IN THE COMPLEMENT-FIXATION TEST:
    A. Amount of complement is not important
    B. Sensitized red cells are added to determine the presence or absence of fixation of complement
    C. Nature of the antigen is important
    D. Serum control is not necessary
    E. The use of a control specimen is not necessary
    Ref. 10 - p. 405

246. THE DIRECT COOMBS' TEST:
    A. Is of little clinical significance
    B. Is performed by immunizing a rabbit with human gamma globulin
    C. Utilizes serum of rabbits immunized with red cells of Rhesus monkeys
    D. Is useful in demonstrating nonagglutinating antibody of infants with erythroblastosis
    E. Is a test for direct-reacting bilirubin
    Ref. 10 - p. 374

247. A CHILD WAS FOUND TO BE UNABLE TO RESPOND IMMUNOLOGICALLY TO PNEUMOCOCCAL ANTIGENS, TYPHOID VACCINE, AND DIPHTHERIA TOXOID AND HE DID NOT DEVELOP THE USUAL ANTIBODIES AFTER RECOVERY FROM MUMPS. WHEN HIS SERUM WAS SUBJECTED TO ELECTROPHORESIS, IT WAS FOUND TO BE ALMOST COMPLETELY DEFICIENT IN:
    A. Albumin
    B. Beta globulin
    C. Gamma globulin
    D. Alpha$_1$ globulin
    E. Alpha$_2$ globulin
    Ref. 7 - p. 79

SECTION II - IMMUNOLOGY

248. A UNIQUE CLASS OF ANTIBODY INVOLVED IN IMMEDIATE-TYPE HYPERSENSITIVITY IN MAN:
   A. γ A
   B. γ M
   C. γ G
   D. γ E
   E. γ D
   Ref. 16 - p. 65

249. ANAPHYLAXIS:
   A. Is the most rapid hypersensitivity reaction of the immediate-type
   B. May be either systemic or localized
   C. Is characterized by an explosive response occurring within minutes of the challenging dose
   D. All of the above
   E. None of the above
   Ref. 16 - p. 185

250. OF THE MANY COMPONENTS OF COMPLEMENT, THE LEVEL OF THE FOLLOWING COMPONENT IS 10-FOLD GREATER THAN ANY OTHER COMPONENT:
   A. C1
   B. C2
   C. C3
   D. C4
   E. C5
   Ref. 16 - p. 122

251. THE LYMPHATIC TISSUE THAT RESPONDS ACTIVELY TO ANTIGENIC STIMULATION:
   A. Lymph nodules
   B. Lymph nodes
   C. Spleen
   D. All of the above
   E. None of the above
   Ref. 16 - p. 29

252. THE FOLLOWING CLASS OF IMMUNOGLOBULIN IS OFTEN REFERRED TO AS HOMOCYTOTROPIC ANTIBODY:
   A. γ G
   B. γ A
   C. γ M
   D. γ D
   E. γ E
   Ref. 16 - p. 103

253. THE SECONDARY MANIFESTATIONS OF THE ANTIGEN-ANTIBODY REACTION INCLUDE:
   A. Agglutination
   B. Precipitation
   C. Neutralization
   D. Complement-dependent reactions
   E. All of the above
   Ref. 16 - p. 132

254. THE CLINICAL MANIFESTATIONS OF HYPERACTIVITY OF CELLULAR IMMUNITY (DELAYED HYPERSENSITIVITY) INCLUDE:
   A. Drug allergies
   B. Certain auto-immune diseases
   C. Contact hypersensitivity
   D. All of the above
   E. None of the above
   Ref. 16 - p. 150

255. WHEN PERITONEAL EXUDATE CELLS FROM SENSITIZED ANIMALS ARE PUT INTO A CAPILLARY TUBE, THE MACROPHAGES WHICH NORMALLY MIGRATE PERIPHERALLY FROM THE END OF THE TUBE ARE INHIBITED IN THE PRESENCE OF THE SENSITIZING ANTIGEN. THIS PHENOMENON IS:
   A. Due to a mitogenic factor
   B. Species-specific
   C. Due to the migration inhibitory factor (MIF)
   D. Due to a specific antibody
   E. All of the above
   Ref. 16 - p. 152

SECTION II - IMMUNOLOGY

256. TOLERANCE:
 A. Is antigen-mediated
 B. May occur as a natural or as an acquired event
 C. May be permanent or transient
 D. All of the above
 E. None of the above
           Ref. 16 - p. 161

257. CONTACT DERMATITIS IS USUALLY CAUSED BY:
 A. Plants (poison ivy)
 B. Cosmetics
 C. Detergents
 D. All of the above
 E. None of the above
           Ref. 16 - p. 349

258. FOLLOWING THE IN VITRO INTERACTION OF THE SENSITIZED LYMPHOCYTE WITH ITS SPECIFIC ANTIGEN, A SUBSTANCE IS RELEASED WHICH HAS THE CAPACITY TO TRANSFER DELAYED HYPERSENSITIVITY TO ANOTHER NONREACTIVE INDIVIDUAL. THIS IS REFERRED TO AS:
 A. Migration inhibition factor
 B. Transfer factor
 C. Mitogenic factor
 D. Skin-reactive factor
 E. None of the above
           Ref. 16 - p. 152

259. RECENTLY, AS ANTIGEN WAS DETECTED IN THE SERA OF PATIENTS WITH HEPATITIS KNOWN AS:
 A. HL-A antigen
 B. Australia (Au/SH) antigen
 C. Platelet (Pl) antigen
 D. Neutrophil (NA) antigen
 E. None of the above
           Ref. 16 - p. 423

260. IN HUMAN SERUM, IgM ANTIBODIES ARE REPRESENTED BY THE FOLLOWING:
 A. Forssman
 B. Wasserman
 C. Cold agglutinins
 D. All of the above
 E. None of the above
           Ref. 13 - p. 31

261. THE FOLLOWING TYPE OF SENSITIVITY IS CONSIDERED BY SOME NOT TO BE A TRUE IMMEDIATE SENSITIVITY INASMUCH AS IT TAKES AT LEAST 6 TO 12 HOURS TO REACH A MAXIMUM SENSITIVITY REACTION:
 A. Prausnitz-Küstner (PK)
 B. Acute anaphylaxis
 C. Passive cutaneous anaphylaxis
 D. Arthus sensitivity
 E. None of the above
           Ref. 13 - p. 155

262. ANTIBODIES THAT PARTICIPATE IN SPECIFIC IMMUNITY TO VIRAL DISEASES INCLUDE:
 A. IgG
 B. IgM
 C. IgA
 D. All of the above
 E. None of the above
           Ref. 13 - p. 325

263. THE FOLLOWING CLASS OF ANTIBODY IS PRIMARILY STIMULATED IN THE CASE OF HIGHLY PRODUCTIVE ACUTE TYPES OF BACTERIAL INFECTION:
 A. γM
 B. γG
 C. γA
 D. All of the above
 E. None of the above
           Ref. 16 - p. 261

SECTION II - IMMUNOLOGY

264. THE INHIBITION OF REPRODUCTION OF <u>TRYPANOSOMA LEWISI</u> WHICH OCCURS DURING THE FIRST FEW DAYS OF INFECTION IN RATS IS DUE TO THE ACQUISITION OF A HUMORAL SUBSTANCE WHICH TALIAFERRO HAS NAMED:
    A. Properdin
    B. Lysozyme
    C. Ablastin
    D. Taliaferro antibody
    E. Opsonin
    Ref. 7 - p. 191

265. ALL OF THE STATEMENTS DESCRIBE THE PROPERDIN SYSTEM, EXCEPT:
    A. It is a normally existing constituent of the serum
    B. Properdin itself has bactericidal and virucidal properties
    C. Properdin is effective when combined with complement and Mg^{++}
    D. It is a euglobulin
    E. It shows no significant difference in level between the newborn and the aged
    Ref. 7 - p. 7

266. REAGINS:
    A. Precipitate the antigen
    B. Can passively sensitize the human skin
    C. Can always neutralize the antigen
    D. Are heat-stable
    E. None of the above
    Ref. 10 - p. 448

267. IMMUNITY TO DIPHTHERIA DEPENDS UPON:
    A. The development by the individual of an adequate concentration of antitoxin in his circulation
    B. The cellular response in the individual
    C. The development of antibacterial antibodies
    D. All of the above
    E. None of the above
    Ref. 10 - p. 556

268. ANTIVIRAL ANTIBODIES MAY BE TESTED FOR BY:
    A. Inhibition of hemagglutination
    B. Precipitation
    C. Complement fixation
    D. Neutralization
    E. All of the above
    Ref. 10 - p. 563

269. IF AN INDIVIDUAL WHO RECEIVED THE SCHICK TEST DOSE ON THE LEFT ARM AND THE CONTROL DOSE ON THE RIGHT ARM SHOWED NEGATIVE REACTIONS ON BOTH ARMS, THESE REACTIONS CAN BE INTERPRETED AS:
    A. Not immune, not allergic
    B. Immune, not allergic
    C. Not immune, allergic
    D. Immune, allergic
    E. Invalid
    Ref. 10 - p. 620

270. WHICH IS NOT A CHARACTERISTIC FEATURE OF THE 19S ANTIBODY?
    A. It has a higher molecular weight than the 7S antibody
    B. It contains about 5X as much carbohydrate as the 7S antibody
    C. It has a greater electrophoretic mobility at pH 8.6 than the 7S antibody
    D. The molecule consists of 2 kinds of polypeptide chain, a light and a heavy chain
    E. It is referred to as IgG
    Ref. 10 - pp. 63-64

SECTION II - IMMUNOLOGY

271. WHEN AN ANTIGEN REENTERS THE HOST TISSUE FOLLOWING AN INITIAL CONTACT, THE SPECIFIC ANTIBODY WILL APPEAR AND REACH A GIVEN LEVEL IN A MUCH SHORTER TIME THAN FOLLOWING THE FIRST CONTACT. THIS IS KNOWN AS:
    A. Hypersensitive reaction
    B. Anamnestic reaction
    C. Inhibition reaction
    D. Tolerance
    E. None of the above
    Ref. 10 - p. 93

272. THE RESULTS OF A GEL DIFFUSION TEST MAY GIVE THE FOLLOWING INFORMATION, EXCEPT:
    A. Each antigen-antibody reaction produces a separate line of precipitation
    B. Generally, the number of lines of precipitation is equal to or less than the number of antigen-antibody systems in the reactants
    C. The thickness of the line may provide a crude measure of the quantity of antigen or antibody
    D. The titer or precise assay of concentration of either antigen or antibody
    E. The position of the line may suggest estimates of relative quantities of the ingredients
    Ref. 7 - pp. 158-159

273. LYSIS IN THE PRESENCE OF ANTIBODY AND COMPLEMENT HAS BEEN SEEN WITH:
    A. Cholera vibrio
    B. Typhoid bacillus
    C. Paratyphoid bacilli
    D. Dysentery bacilli
    E. All of the above
    Ref. 7 - p. 162

274. THE STATEMENTS DESCRIBE SHWARTZMANN-SANARELLI PHENOMENON, EXCEPT:
    A. The provocative injection must be given intravascularly
    B. The provocative injection must follow within a limited interval of time after the preparatory injection
    C. Bacterial and other substances may induce reactivity and elicit the reaction
    D. It is of hypersensitive nature
    E. There is no antigenic specificity involved
    Ref. 7 - p. 370

275. WHICH OF THE FOLLOWING STATEMENTS DOES NOT APPLY TO TISSUE GRAFTING?:
    A. Grafted skin is usually sloughed at about 10 to 14 days after its first application to a random homologous recipient
    B. If tissue from the same donor is then reapplied to the same recipient, the rejection occurs much more quickly
    C. Graft rejection reveals the specificity characteristic of immunologic phenomena
    D. Fetal animals injected with cells of a potential graft donor are rendered tolerant to transplants in postnatal life
    E. The rejection state cannot be transferred from an immune subject to a homologous normal recipient by means of lymphoid cells
    Ref. 7 - p. 551

276. ANTIBODY-LIGAND INTERACTIONS ARE INFLUENCED BY VARIATIONS IN:
    A. pH
    B. Temperature
    C. Ionic strength
    D. All of the above
    E. None of the above
    Ref. 12 - p. 372

SECTION II - IMMUNOLOGY

277. IF A SOLUTION OF PURE ANTIGEN IS PLACED IN TWO ADJACENT WELLS AND THE HOMOLOGOUS ANTIBODY IS PLACED IN THE CENTER WELL, THE TWO PRECIPITIN BANDS JOIN AT THEIR CONTIGUOUS ENDS AND FUSE. THIS PATTERN IS KNOWN AS:
    A. Reaction of identity
    B. Reaction of nonidentity
    C. Reaction of partial identity
    D. Cross-reaction
    E. None of the above
    Ref. 12 - p. 399

278. OF THE FOLLOWING METHODS USED AS ROUTINE ASSAYS FOR MEASURING ANTIBODY ACTIVITY, THE MOST SENSITIVE METHOD IS THE:
    A. Bacterial agglutination
    B. Passive hemagglutination
    C. Precipitin reactions in liquid media
    D. Precipitin reactions in agar gel
    E. Complement fixation
    Ref. 12 - p. 407

279. IN MOST NORMAL AND HYPERIMMUNE INDIVIDUALS, WHAT PERCENTAGE OF THE IMMUNOGLOBULINS ARE γG PROTEINS?:
    A. 10 percent
    B. 25 percent
    C. 50 percent
    D. 75 percent
    E. Over 85 percent
    Ref. 12 - p. 417

280. THE ALLERGY-MEDIATING ANTIBODIES IN MAN REPRESENT A SPECIAL CLASS OF IMMUNOGLOBULINS CALLED:
    A. γG
    B. γM
    C. γA
    D. γE
    E. γD
    Ref. 12 - p. 433

281. THE MOLECULAR WEIGHT OF γG IMMUNOGLOBULINS IS CLOSE TO:
    A. 10,000
    B. 50,000
    C. 150,000
    D. Unknown
    E. None of the above
    Ref. 12 - p. 418

282. THE COMPLEMENT SYSTEM CONSISTS OF A GROUP OF THE FOLLOWING NUMBER OF PROTEINS WHICH NORMALLY EXIST IN SERUM:
    A. 4
    B. 5
    C. 6
    D. 11
    E. More than 12
    Ref. 12 - p. 523

283. IN GENERAL, AFTER THE INJECTION OF THE IMMUNOGEN, γM ANTIBODIES ARE USUALLY DETECTABLE:
    A. Within the first week
    B. After 1 to 2 weeks
    C. After 3 to 4 weeks
    D. After 1 month
    E. None of the above
    Ref. 12 - p. 463

284. THE CLONAL SELECTION THEORY OF ANTIBODY FORMATION WAS PROPOSED BY:
    A. Ehrlich
    B. Haurowitz
    C. Mudd
    D. Burnet
    E. Taliaferro
    Ref. 12 - p. 502

285. TRANSPLANTS FROM ONE REGION TO ANOTHER OF THE SAME INDIVIDUAL ARE REFERRED TO AS:
    A. Isografts
    B. Autografts
    C. Homografts
    D. Allografts
    E. Heterografts
    Ref. 12 - p. 591

SECTION II - IMMUNOLOGY   41

ANSWER THE FOLLOWING QUESTIONS BY USING THE KEY
OUTLINED BELOW:
A. If only A is correct
B. If only B is correct
C. If both A and B are correct
D. If neither A nor B is correct

286. THE CLASSICAL DELAYED HYPERSENSITIVITY AND THE JONES-
MOTE TYPE HAVE IN COMMON:
A. Reactivity appears in the absence of serum antibodies
B. The state may be transferred by leukocytes
C. Both
D. Neither   Ref. 7 - pp. 349, 350

287. UNIVALENT ANTIBODY IS AN ANTIBODY WITH:
A. A single complete pattern against the major determinant
B. Two or more patterns against the major determinant
C. Both
D. Neither   Ref. 7 - p. 96

288. THE CAPACITY TO STIMULATE THE FORMATION OF ANTIBODIES:
A. Immunogenicity
B. Specificity
C. Both
D. Neither   Ref. 12 - p. 359

289. THE REACTION RESULTING FROM BACTERIAL CELLS SUSPENDED
IN SERUM FROM AN ANIMAL PREVIOUSLY INJECTED WITH THESE
BACTERIA:
A. Precipitation
B. Agglutination
C. Both
D. Neither   Ref. 12 - p. 359

290. ANTIBODIES ARE:
A. Proteins
B. Immunoglobulins
C. Both
D. Neither   Ref. 12 - p. 360

291. HAPTENS:
A. Are not immunogenic
B. React selectively with antibodies of the appropriate specificity
C. Both
D. Neither   Ref. 12 - p. 361

292. PROTEINS BEARING SUBSTITUENTS THAT ARE COVALENTLY
LINKED TO THEIR AMINO ACID SIDE CHAINS ARE USUALLY
REFERRED TO AS:
A. Haptenic groups
B. Antigenic determinants
C. Both
D. Neither   Ref. 12 - p. 361

SECTION II - IMMUNOLOGY

293. IN THE EQUIVALENCE ZONE OF A PRECIPITIN REACTION, THE SUPERNATANTS ARE USUALLY DEVOID OF:
   A. Detectable antibody
   B. Detectable antigen
   C. Both
   D. Neither
   Ref. 12 - p. 380

294. THE REACTIONS OF AN ANTISERUM WITH HETEROLOGOUS ANTIGENS ARE CALLED:
   A. Passive agglutination reactions
   B. Cross-reactions
   C. Both
   D. Neither
   Ref. 12 - p. 391

295. REMOVAL OF CERTAIN CROSS-REACTING ANTIBODIES IS KNOWN AS:
   A. Adsorption
   B. Avidity
   C. Both
   D. Neither
   Ref. 12 - p. 395

296. THE PRECIPITIN REACTION CAN BE CARRIED OUT IN:
   A. Gels
   B. Liquid systems
   C. Both
   D. Neither
   Ref. 12 - p. 397

297. THE DOUBLE DIFFUSION IN TWO DIMENSIONS OF PLACING ANTIGEN AND ANTIBODY SOLUTIONS IN SEPARATE WELLS CUT IN AN AGAR PLATE WAS DEVELOPED BY:
   A. Oudin
   B. Ouchterlony
   C. Both
   D. Neither
   Ref. 12 - p. 399

298. GRABAR AND WILLIAMS DEVELOPED AN AGAR GEL METHOD FOR IDENTIFYING ANTIGENS IN COMPLEX MIXTURES KNOWN AS:
   A. Immunoelectrophoresis
   B. Immunofluorescence
   C. Both
   D. Neither
   Ref. 12 - p. 401

299. IMMUNOELECTROPHORESIS COMBINES THE PRINCIPLES OF:
   A. Electrophoresis
   B. Precipitation
   C. Both
   D. Neither
   Ref. 12 - p. 401

300. IN AN AGGLUTINATION TITRATION TEST, IF A 1:512 DILUTION GIVES PERCEPTIBLE AGGLUTINATION BUT A 1:1024 DILUTION DOES NOT, THE TITER IS:
   A. 512
   B. 1024
   C. Both
   D. Neither
   Ref. 12 - p. 403

301. IN THE AGGLUTINATION ASSAY, THE ONLY ANTIBODIES THAT CAN CAUSE AGGLUTINATION ARE THOSE SPECIFIC FOR:
   A. Surface determinants
   B. Internal components
   C. Both
   D. Neither
   Ref. 12 - p. 404

302. IN THE AGGLUTINATION REACTION, WHEN UNDILUTED OR ONLY SLIGHTLY DILUTED SERUM DOES NOT VISIBLY REACT WITH THE ANTIGEN PARTICLES, THE REGION IS CALLED THE:
   A. Prozone
   B. Postzone
   C. Both
   D. Neither
   Ref. 12 - p. 404

SECTION II - IMMUNOLOGY

303. THE INHIBITORY ANTIBODY MOLECULES IN AGGLUTINATION ARE REFERRED TO AS:
    A. Blocking antibodies
    B. Incomplete antibodies
    C. Both
    D. Neither
    Ref. 12 - p. 404

304. INCOMPLETE ANTIBODY CAN BE DETECTED BY BINDING TO AN ANTIGEN ON THE RED CELL SURFACE AND IN COMPLEXING WITH ANTIBODY TO ITSELF. THE TEST IS KNOWN AS THE:
    A. Antiglobulin test
    B. Coombs' test
    C. Both
    D. Neither
    Ref. 12 - p. 405

305. THE PRESENCE OF A HIGH TITER OF ANTIBODIES, WITHOUT A CHANGE OVER A PERIOD OF A FEW WEEKS, TO A GIVEN INFECTIOUS AGENT MAY INDICATE:
    A. Past infection
    B. Vaccination
    C. Both
    D. Neither
    Ref. 12 - p. 410

306. FOR USE IN THE ELECTRON MICROSCOPE, ANTIBODIES ARE RENDERED HIGHLY ELECTRON-SCATTERING BY ATTACHING A MOLECULE OF THE FOLLOWING MOLECULE:
    A. Ferritin
    B. Fluorescein isothiocyanate
    C. Both
    D. Neither
    Ref. 12 - p. 413

307. THE FOLLOWING PROTEIN PASSES ACROSS THE PLACENTA FROM MATERNAL TO FETAL CIRCULATION IN MAN:
    A. γA
    B. γM
    C. Both
    D. Neither
    Ref. 12 - p. 432

308. THE NORMAL IMMUNOGLOBULIN POPULATION IN ANY PERSON INCLUDES:
    A. k-chain
    B. λ-chain
    C. Both
    D. Neither
    Ref. 12 - p. 436

309. IN ORDER TO INDUCE SYNTHESIS OF ANTIBODIES, A SUBSTANCE MUST ORDINARILY BE RECOGNIZED AS:
    A. "Self"
    B. Alien
    C. Both
    D. Neither
    Ref. 12 - p. 454

310. HIGHLY IMMUNOGENIC SUBSTANCES INCLUDE:
    A. Bacterial toxins
    B. Immunoglobulins
    C. Both
    D. Neither
    Ref. 12 - p. 457

311. ANTIBODY FORMATION IS STIMULATED BY THE PURIFIED FORMS OF:
    A. Many polysaccharides
    B. All nucleic acids
    C. Both
    D. Neither
    Ref. 12 - p. 455

312. PASSIVE TRANSFER OF MATERNAL ANTIBODIES TO NEWBORN IS ACHIEVED IN MAMMALS BY:
    A. Transfer from maternal to fetal circulation in utero
    B. Suckling of colostrum
    C. Both
    D. Neither
    Ref. 12 - p. 496

SECTION II - IMMUNOLOGY

313. THE FUNCTION OF ADJUVANTS HAS BEEN CONSIDERED:
    A. To increase surface area of antigen
    B. To prolong retention of antigen
    C. Both
    D. Neither   Ref. 16 - p. 100

314. IN THE GUINEA PIG, PCA REACTIONS ARE BELIEVED TO BE PRODUCED BY ANTIBODIES OF THE:
    A. $7S\gamma_1$ type
    B. $7S\gamma_2$ type
    C. Both
    D. Neither   Ref. 16 - p. 189

315. THE ARTHUS REACTION MAY BE ELICITED AS FOLLOWS:
    A. Active Arthus reaction
    B. Direct passive Arthus reaction
    C. Both
    D. Neither   Ref. 16 - p. 200

ANSWER THE FOLLOWING QUESTIONS BY USING THE KEY OUTLINED BELOW:
A. If both statement and reason are true and related cause and effect
B. If both statement and reason are true but not related cause and effect
C. If the statement is true but the reason is false
D. If the statement is false but the reason is true
E. If both statement and reason are false

316. Precipitating antibody is essential for the Arthus reaction BECAUSE the reaction depends chiefly upon injury to blood vessels.
                                                  Ref. 10 - pp. 438, 439

317. The term immunogen has been proposed to define any substance capable of evoking an immune response BECAUSE certain chemical groupings on the immunogen which determine specificity of the immunologic reaction are referred to as determinant groups.
                                                  Ref. 16 - p. 100

318. The heavy-chain types in all classes of immunoglobulins are called (k) and lambda (λ) BECAUSE the light-chain types are responsible for the observed biological differences between the various classes.
                                                  Ref. 16 - p. 104

319. Immunologic adjuvants are substances which, if mixed with antigens, cause an increase in antibody responses BECAUSE the adjuvants themselves are antigenic.   Ref. 7 - p. 124

320. Antibody formation is limited to intact organisms BECAUSE the organs of immunized animals are not able to continue production of antibodies in vitro.   Ref. 10 - pp. 88-89

321. Antigenicity is not restricted to substances produced by microorganisms and parasites BECAUSE protein poisons such as snake venom may cause the production of neutralizing antibodies.
                                                  Ref. 10 - p. 117

322. In vitro antigen-antibody reactions are carried out at a specific temperature in buffered media containing electrolytes BECAUSE the ease of combination is dependent upon these factors.
                                                  Ref. 16 - pp. 135-136

SECTION II - IMMUNOLOGY

323. To induce delayed hypersensitivity with contactants, they need not be able to bind covalently to proteins BECAUSE such materials are usually immunogenic and do not require carrier proteins.
Ref. 16 - p. 220

324. In systemic infectious diseases the primary immunity mechanism appears to be effected by serum antibody of the γA variety BECAUSE they involve a complex pathogenesis with passage of the organisms or toxins through the blood stream.
Ref. 16 - p. 488

325. Live virus immunization procedures are delayed until the end of the first year of life BECAUSE the passively acquired antibody can inhibit successful immunization of infants with parenteral live virus vaccines.
Ref. 16 - p. 493

326. An antibody molecule cannot serve as an antigen BECAUSE it does not possess surface sites which serve as antigen-determinants.
Ref. 13 - p. 28

327. Tolerance is produced more readily in embryonic life than later in life BECAUSE inducible cells in the embryo are either more easily repressed by antigen or are less numerous than in older animals.
Ref. 13 - p. 90

328. The BCG vaccine for tuberculosis utilizes killed bacteria BECAUSE killed organisms are superior to living organisms in affording protection.
Ref. 7 - pp. 230-231

329. A rabies vaccine prepared in embryonated duck eggs and inactivated by beta propriolactone is said to be comparable in its antibody-inducing effect to the traditional nerve tissue vaccine BECAUSE both are killed vaccines.
Ref. 7 - p. 230

330. Infants are protected against various infectious diseases, such as diphtheria, measles, and mumps BECAUSE antibodies can be transmitted from mother to fetus through the placenta.
Ref. 7 - p. 234

331. Anaphylaxis is generally regarded as the prototype of hypersensitivities of the immediate type BECAUSE anaphylactic shock depends upon the reactions of certain types of tissue, the so-called shock tissues.
Ref. 7 - pp. 293, 294

332. Much has been devoted to the study of anaphylaxis in the dog BECAUSE this animal readily becomes sensitized and regularly develops acute and characteristic shock upon later reinjection of antigen.
Ref. 7 - p. 295

333. The Arthus reaction apparently stems directly from the precipitative union of antigen with antibody, presumably BECAUSE the precipitate has no irritative effect upon vascular tissues.
Ref. 7 - p. 282

334. Hypersensitivity to tuberculin has been transferred from sensitized donors to normal recipients via cells BECAUSE cells of the lymphoid series are responsible for establishing the reactive state.
Ref. 7 - p. 340

335. The vaccine used for smallpox is not entirely without virulence BECAUSE generalized vaccinial pox may occur, and very rarely encephalitis may follow vaccination.
Ref. 7 - p. 231

SECTION II - IMMUNOLOGY

336. The conditions favorable for establishing the anaphylactic state are in general those which are favorable for the induction of antibody formation BECAUSE in both cases only very small doses of antigen are required.
Ref. 7 - p. 293

337. Evanescent cutaneous reaction is identical to the Arthus reaction BECAUSE both reactions become visible within a few minutes and disappear within three or four hours, and they never progress to necrosis.
Ref. 7 - p. 292

338. Immune serum for the treatment of certain bacterial diseases, such as those caused by pneumococcus and meningococcus, is no longer used BECAUSE antibiotics are now employed for the treatment of diseases caused by these organisms. Ref. 7 - p. 235

339. Passive immunization is no longer employed prophylactically for bacterial diseases BECAUSE protection from passive transfer of antibodies is inferior to active immunization. Ref. 7 - p. 235

340. All appropriate combinations of antibody and antigen succeed in releasing histamine BECAUSE the histamine may cause mainly contractions of smooth muscle and increased permeability of capillaries.
Ref. 7 - p. 289

341. The serum levels of IgG antibodies are critical to the defense of the infant against infectious diseases BECAUSE these antibodies are of major importance in antimicrobial defense. Ref. 13 - p. 103

342. The passive hemagglutination test is often more sensitive than bacterial agglutination BECAUSE of the relatively large size of the antigen particle.
Ref. 2 - p. 368

343. Cow serum is used in conglutination reaction BECAUSE it has the natural antibody for the erythrocytes and the clumping activity.
Ref. 2 - p. 368

344. Delayed hypersentivity differs from anaphylaxis and atopic allergy BECAUSE the delayed type of hypersensitivity is not passively transferable. Ref. 2 - p. 411

345. The Arthus phenomenon is a generalized hypersensitivity reaction produced in actively and passively sensitized animals BECAUSE the rapidity of the reaction is correlated with the amount of precipitable circulating antibody.
Ref. 2 - p. 407

346. The molecular weight of antibodies is expected to be appropriate to globulins BECAUSE antibodies are globulins.
Ref. 10 - p. 61

347. The majority of antibody activity is due to a protein with a sedimentation constant of 7S BECAUSE antibodies are always found in the globulin fraction of serum, never in the albumin.
Ref. 10 - p. 62

348. Antigenicity is restricted to substances produced by microorganisms and parasites BECAUSE a substance is said not to be antigenic unless it will cause the production of antibodies or initiate some other "immune" response, such as sensitization. Ref. 10 - p. 117

SECTION II - IMMUNOLOGY

349. Adjuvants are substances which when mixed with antigens improve antibody production or other immune response BECAUSE they can never be antigenic themselves. Ref. 10 - p. 125

350. The heterogenetic antigens of microorganisms are usually called common antigens BECAUSE they are common to more than one kind of organism. Ref. 2 - p. 347

351. When tissue is transplanted from one part of the body of a mammal to another, it is called an autograft BECAUSE the donor and recipient are the same individual. Ref. 2 - p. 340

352. Forssman antigen is a lipid BECAUSE it can be extracted from tissues containing it with alcohol. Ref. 2 - p. 341

353. Embryonic tissue does not form antibody, apparently BECAUSE it is unable to distinguish the foreign nature of an antigenic substance not normally present in that tissue. Ref. 2 - p. 342

354. The immunoglobulin molecule consists of four peptide chains, two heavy chains, and two light chains, joined by disulfide bonds BECAUSE the molecular weight of IgG agrees well with the sum of the molecular weights of its components. Ref. 2 - p. 349

355. Forssman antigen is known as heterophile antigen BECAUSE it is found in blood rather than in tissues and organs. Ref. 2 - p. 341

356. Nonprecipitating antibody is not necessarily univalent BECAUSE some nonprecipitating antibodies will precipitate with antigen if complement is added. Ref. 12 - p. 387

357. The γG immunoglobulins are often referred to as $7S\gamma$-globulins BECAUSE their sedimentation coefficient in neutral, dilute salt solution is 6.5 to 6.6 Svedberg units. Ref. 12 - p. 418

358. Each γG molecule contains two identical heavy chains and two identical light chains BECAUSE the Gm markers are found on the heavy chains of γG immunoglobulins and the InV markers on light chains. Ref. 12 - pp. 424-425

359. Mercaptan is not used as an assay to distinguish between γM and γG antibodies BECAUSE the treatment inactivates γG molecules. Ref. 12 - p. 431

360. When nucleic acids are injected alone, they do not evoke antibody formation BECAUSE of the possible nuclease activity <u>in vivo</u>. Ref. 12 - p. 455

361. Infectious, but attenuated, agents probably tend to persist for a shorter period than noninfectious substances BECAUSE infectious agents are not self-replicating. Ref. 12 - p. 472

362. The rate of increase in serum antibody concentration is less in the secondary response BECAUSE more antibody-forming cells are present. Ref. 12 - p. 475

SECTION II - IMMUNOLOGY

363. Induction of tolerance in adults is much easier after massive X-irradiation or treatment with immunosuppressive drugs BECAUSE the animals are almost completely depleted of antibody-forming cells.
Ref. 12 - p. 489

364. The principal requirement for the Arthus reaction seems to be the formation of microprecipitates in the tissues BECAUSE nonprecipitating antibodies appear to be ineffective. Ref. 12 - p. 544

365. Primary antibody response affords much less protection than secondary response BECAUSE antibodies appear more slowly and their over-all combining power is relatively poor, at least initially.
Ref. 12 - p. 466

366. Antibodies with low avidity are the most efficient for producing agglutination BECAUSE the binding forces of antibody must overcome the strong repelling forces between antigen particles.
Ref. 13 - p. 111

367. The "prozone" often occurs in an agglutination reaction BECAUSE all of the antigenic sites on the bacterial cell are quickly saturated with antibody in the form of a primary complex.
Ref. 13 - p. 111

368. In the guinea pig, $7S\gamma_1$ antibodies are effective for producing the Arthus reaction BECAUSE γ_1 antibodies are able to fix C'.
Ref. 13 - p. 158

369. The stomach provides an effective barrier against the passage of bacteria from the mouth to the intestines BECAUSE most microbes that reach the stomach are destroyed by the low pH of the gastric contents.
Ref. 13 - p. 310

370. Lymphotoxin is a substance shown to be liberated from specifically sensitized lymphocytes BECAUSE lymphotoxin seems to be associated with target cell injury. Ref. 16 - p. 153

ANSWER THE FOLLOWING QUESTIONS BY USING THE KEY OUTLINED BELOW:
A. If 1 and 4 are correct
B. If 2 and 3 are correct
C. If 1, 2 and 3 are correct
D. If all are correct
E. If all are incorrect
F. If some combination other than the above is correct

371. THE CHARACTERISTICS OF THE ANTIGEN-ANTIBODY REACTION LEADING TO PRECIPITATION:
1. The reaction is highly specific
2. The reaction takes place in two stages
3. The reaction is reversible
4. The reaction is irreversible   Ref. 13 - p. 110

372. PASSIVELY ACQUIRED IMMUNITY IS ACCOMPLISHED BY MEANS OF:
1. Transplacental transfer of maternal γG antibody
2. Transplacental transfer of maternal γM antibody
3. Transfer of antibody in breast milk
4. Passive immunization by serum gamma globulin (immunoprophylaxis)
Ref. 16 - p. 492

SECTION II - IMMUNOLOGY

373. ANTIBODIES ARE SYNTHESIZED IN ABUNDANCE BY LYMPHOID CELLS IN THE:
    1. Spleen
    2. Bone marrow
    3. Lymph nodes
    4. Thymus Ref. 13 - p. 52

374. GASTROINTESTINAL ALLERGY:
    1. Most frequently implicated foods include milk, eggs, citrus fruits, and chocolate
    2. The symptoms are thought to be mediated through precipitating antibody
    3. Skin testing with food antigens is of prime diagnostic importance
    4. It is manifest clinically by the rejection of the foreign antigen through vomiting and diarrhea Ref. 16 - p. 346

375. CYTOPHILIC ANTIBODIES HAVE BEEN REPORTED TO OCCUR IN THE IMMUNOGLOBULIN CLASSES:
    1. IgG
    2. IgM
    3. IgE
    4. IgA Ref. 13 - p. 130

376. A GOOD IMMUNE RESPONSE DEPENDS UPON:
    1. Greater solubility of antigen
    2. Longer period of time that antigen remains in tissues
    3. Use of adjuvant
    4. Rapid dispersion of antigen in the tissues Ref. 2 - p. 336

377. PROPERTIES ASSOCIATED WITH ANTIGENICITY:
    1. Substances of natural origin of relatively large molecular size
    2. Complete antigens are usually naturally occurring proteins containing a full complement of amino acids
    3. They are foreign, or contain structures foreign, to the antibody producing mechanism
    4. They are usually polysaccharides of low molecular weight Ref. 2 - p. 337

378. ANTIGENICITY DECREASES:
    1. With low molecular weight
    2. With enzymatic hydrolysis
    3. When derived proteins, such as gelatin, are in conjugated form
    4. With high molecular weight Ref. 2 - p. 337

379. THE FOLLOWING MAY ACT AS HAPTENS:
    1. Polysaccharides
    2. Lipids
    3. Drugs
    4. Cutting oils Ref. 2 - pp. 344-345

380. ANTIBODY ACTIVITY IS A PROPERTY OF MORE THAN ONE MOLECULAR SPECIES, WHICH ARE SEPARABLE BY:
    1. Electrophoresis
    2. Sedimentation
    3. Immunoelectrophoresis
    4. Diffusion rates Ref. 2 - p. 348

SECTION II - IMMUNOLOGY

381. GAMMA GLOBULIN FRACTION OF GLOBULIN:
   1. Contains small amounts of albumin
   2. Contains most of the antibodies
   3. Migrates slowest of all the three globulin fractions in electrophoresis
   4. Does not contain separable parts   Ref. 2 - p. 347

382. AFTER THE ANTIGEN AND ANTIBODY HAVE REACTED, THEY ARE SPECIFICALLY ADSORBED TO ONE ANOTHER AND HELD TOGETHER BY:
   1. Electrostatic attraction between oppositely charged groups
   2. Van der Waals forces
   3. Dipolar interaction between polar but dissociated groups
   4. Covalent bonds   Ref. 2 - p. 357

383. VALENCE:
   1. Refers to a reaction site of antigen or antibody
   2. Of blocking antibody is monovalent
   3. Of complete antigens is multivalent
   4. Of simple haptens is monovalent   Ref. 2 - pp. 357-358

384. PROZONE PHENOMENON:
   1. Is the lack of manifestation of antigen-antibody reaction in the zone of antibody excess
   2. Refers to the failure to form a precipitate due to antigen excess
   3. Is observed when the antigen-antibody ratio is at its optimal proportion
   4. May be due to insufficient antigen
      Ref. 2 - p. 359

385. THE SECOND STAGE OF ANTIGEN-ANTIBODY REACTION OCCURS IN:
   1. Precipitation
   2. Agglutination
   3. Neutralization of toxin
   4. Lysis   Ref. 2 - p. 364

386. PRECIPITATION AS A MANIFESTATION OF ANTIGEN-ANTIBODY RESPONSE:
   1. Depends on presence of electrolytes
   2. Is quantitatively affected by atmospheric pressure
   3. Is speeded by opsonins
   4. Does not require heat labile substance such as complement
      Ref. 2 - p. 365

387. THE EQUIVALENCE ZONE:
   1. Is that proportion of antigen and antibody such that neither is in excess
   2. May be demonstrated by application of the precipitin ring test to the supernatant
   3. Is the zone in which the maximum amount of precipitate is formed
   4. Precipitate formed at this zone may be used for nitrogen determination
      Ref. 2 - p. 365

388. PRECIPITIN REACTION:
   1. Gel diffusion method of Oudin cannot be used
   2. Nature of antigen is not important
   3. The test is usually carried out by adding increasing amounts of antiserum to a constant amount of antigen
   4. May occur with only complete antigen but not with hapten
      Ref. 2 - pp. 364-367

SECTION II - IMMUNOLOGY

389. PRECIPITIN REACTION:
   1. Is usually used for forensic purposes
   2. Cross-reactions are not observed
   3. Is not highly specific
   4. Is highly specific but cross-reactions occur between similar antigens
   Ref. 2 - p. 367

390. AGGLUTINATION REACTION:
   1. Microorganisms become immobilized in the presence of immune serum
   2. Living bacteria need not be used for dead bacteria are agglutinated as readily
   3. In the presence of homologous antibody bacteria aggregate to form large clumps
   4. Bacteria are killed in the presence of immune serum
   Ref. 2 - p. 367

391. AGGLUTINATION:
   1. The reaction can be used as a slide test
   2. Requires complement
   3. Only bacteria can be used as the antigen
   4. Varying dilutions of serum are added to bacterial suspension and incubated at 37° C overnight or at 55° C for two hours
   Ref. 2 - pp. 367-368

392. CONGLUTINATION REACTION:
   1. May be used as an indicator system like sheep cell-hemolysin system of C-F reaction
   2. Presence or absence of free complement is indicated by conglutination
   3. Presence or absence of free complement is not indicated by conglutination-inhibition
   4. Conglutination is identical to agglutination in that both reactions require complement
   Ref. 2 - p. 368

393. THE FOLLOWING CLASS OF IMMUNOGLOBULIN HAS A KNOWN FUNCTION IN VIRUS-HOST (MAN) INTERACTIONS:
   1. γG
   2. γM
   3. γA
   4. γD
   Ref. 16 - p. 282

394. TOXIN-ANTITOXIN REACTION:
   1. The complete destruction of either component is not necessary
   2. The toxin is neutralized as long as the toxin-antitoxin union persists
   3. Exhibits the Danysz phenomenon
   4. The avidity of an antitoxin for its corresponding toxin is the same in different cases, e.g., diphtheria and tetanus
   Ref. 2 - pp. 373-374

395. COMPLEMENT:
   1. Is heat labile
   2. Is present in normal serum
   3. Does not increase with immunization
   4. Participates in lytic reaction
   Ref. 2 - p. 377

SECTION II - IMMUNOLOGY

396. COMPLEMENT FIXATION:
 1. Specificity is not due to the initial reaction between antigen and antibody
 2. The indicator system is specific
 3. If complement is available for the indicator system, it is a positive test
 4. If complement is not available for the indicator system, it is a negative test   Ref. 2 - p. 380

397. IN ORDER TO OBTAIN THE OPSONIC INDEX THE FOLLOWING ARE NEEDED:
 1. Sheep red cells
 2. Normal and immune sera
 3. Bacteria
 4. Hemolysin   Ref. 2 - p. 382

398. ACTIVE IMMUNITY MAY BE ACQUIRED BY:
 1. Exposure to the infectious disease
 2. Inoculation with living organisms
 3. Inoculation with organisms killed by heat
 4. Inoculation with immune serum   Ref. 2 - p. 399

399. PASSIVE IMMUNITY:
 1. Involves active generation of protective substance by the immunized individual
 2. Is exemplified by the transfer of antibodies from mother to fetus in utero via the placenta
 3. May be brought about by the injection of convalescent serum
 4. Is of long duration   Ref. 2 - p. 403

400. FACTORS WHICH INCREASE PHAGOCYTOSIS INCLUDE:
 1. Neutral or slightly acid pH
 2. Bacteria of low virulence
 3. Presence of immune opsonin
 4. Presence of bacteriophage   Ref. 2 - p. 382

401. IN CELLULAR TRANSFER OF DELAYED HYPERSENSITIVITY, THE FOLLOWING CELLS ARE USUALLY EMPLOYED:
 1. Peritoneal exudate cells
 2. Lymph node cells
 3. Splenic cells
 4. Red blood cells   Ref. 7 - p. 340

402. THE PROPERTIES OF γG ANTIBODIES:
 1. Heat stable
 2. Heat labile
 3. 8S
 4. Highest concentrations in serum   Ref. 16 - p. 189

403. THE PROPERTIES OF γM ANTIBODIES:
 1. 7S
 2. 19S
 3. Molecular weight: 890,000
 4. Molecular weight: 150,000   Ref. 16 - p. 102

SECTION II - IMMUNOLOGY

404. THE FOLLOWING LIVE VACCINES ARE RECOMMENDED FOR INFANTS AND CHILDREN:
   1. Rubeola
   2. Poliomyelitis (Sabin)
   3. Rubella
   4. Mumps
   Ref. 16 - p. 495

405. ANTI-Rh ANTIBODY:
   1. First found in human serum was an agglutinin active against RBCs suspended in saline
   2. It can be absorbed by red cells
   3. It can be eluted from the red cells
   4. It is not active at 37º C
   Ref. 10 - p. 284

406. DANYSZ PHENOMENON:
   1. When toxin is added to antitoxin, the toxicity of the mixture depends partly on how the toxin is added
   2. If an equivalent amount of toxin is added all at once, the mixture is toxic
   3. If an equivalent amount of toxin is added at intervals, in fractions, the final mixture is generally nontoxic
   4. This phenomenon shows the ability of toxin to combine with antitoxin in multiple proportions
   Ref. 10 - p. 341

407. PROPERTIES OF γA GLOBULINS:
   1. Found in external secretions of the body
   2. 19S
   3. Produced early in the immune response
   4. Important in protection against viruses
   Ref. 16 - p. 283

408. CONCERNING THE MECHANISM OF ANAPHYLAXIS:
   1. It is essentially the result of antibody-antigen reaction
   2. Only antigenic or haptenic substances will induce anaphylaxis
   3. Sensitivity can be passively conferred on a normal animal by transfer of serum from a sensitized animal
   4. Antibody can be demonstrated in the serum of sensitive animals
   Ref. 10 - p. 434

409. ARTHUS REACTION:
   1. It depends chiefly upon injury to blood vessels
   2. It is identical with the wheal and flare reaction
   3. It is characterized by edema with extensive cellular filtration, hemorrhage, and secondary necrosis
   4. It is not identical with anaphylactic shock
   Ref. 10 - p. 439

410. ATOPY:
   1. It is usually naturally acquired
   2. Duration of sensitization is long
   3. Predisposition can be inherited from either mother or father
   4. It is induced only by protein
   Ref. 10 - p. 445

411. THE MAJOR CLASSES OF IMMUNOGLOBULINS THAT ARE NOW RECOGNIZED ON THE BASIS OF THEIR PHYSICAL, CHEMICAL, AND ANTIGENIC PROPERTIES ARE:
   1. γ G
   2. γ A
   3. γ M
   4. γ D
   Ref. 12 - p. 417

SECTION II - IMMUNOLOGY

412. THE 19S ANTIBODIES ARE:
  1. Detected early in most immune responses
  2. Detected late in most immune responses
  3. Characterized by lower molecular weight than other immunoglobulins
  4. Characterized by higher molecular weight than other immunoglobulins
  Ref. 12 - p. 431

413. MOST ADJUVANTS SEEM TO HAVE IN COMMON THE FOLLOWING:
  1. They accelerate the destruction of antigen and permit relatively prolonged maintenance of low but effective levels in the tissue
  2. They retard the destruction of antigen and permit relatively prolonged maintenance of low but effective levels in the tissue
  3. They provoke inflammatory responses, the cells of which may play a role in antibody formation
  4. They provoke inflammatory responses, the cells of which do not play a role in antibody formation   Ref. 12 - p. 459

414. AS COMPARED TO THE PRIMARY ANTIBODY RESPONSE, SECONDARY RESPONSE IS CHARACTERIZED BY:
  1. Lowering of the threshold dose of immunogen
  2. Shortening of the lag phase
  3. A higher rate
  4. Longer persistence of antibody synthesis
  Ref. 12 - p. 465

415. IN IMMEDIATE-TYPE HYPERSENSITIVITY, THE STATE OF SENSITIVITY IS TRANSFERRED PASSIVELY BY:
  1. Lymphocytes
  2. Monocytes
  3. Macrophages
  4. Serum   Ref. 15 - p. 152

ANSWER THE FOLLOWING QUESTIONS (T)RUE OR (F)ALSE:

416. The specificity of an antibody appears to be determined by the primary amino acid sequence of the antibody molecule.
  Ref. 16 - p. 106

417. The fragment which contains the antigenic determinants of the γG globulin molecule has been designated the Fc fragment while the two fragments which retain the ability to combine with antigen have been designated the Fab fragments.   Ref. 16 - p. 108

418. The coating of foreign particles and infective agents with the first four components of complement renders the particles immediately susceptible to phagocytosis.   Ref. 16 - p. 129

419. Antilymphocyte serum (ALS) used for immunosuppression is an example of specific immunosuppressive agent.   Ref. 16 - p. 168

420. Acute serum sickness results from immune complexes formed when antibodies are produced against the foreign serum proteins which are still present in the blood stream.   Ref. 16 - p. 206

421. If a hapten-carrier complex is involved in delayed hypersensitivity, the specificity tends to be directed toward the carrier proteins as well as to the hapten.   Ref. 16 - p. 223

SECTION II - IMMUNOLOGY 55

422. There is no in utero transfer of γM associated antibodies, such as those to the gram-negative bacteria, whereas there is passive transfer of G associated antibodies to many viruses and bacteria.
Ref. 16 - p. 249

423. Cell-mediated immunity appears to be of no consequence in the defense mechanism of most types of fungal infections.
Ref. 16 - p. 321

424. The fundamental role of cell-mediated immunity in tumor immunity is the ability to transfer a tumor immune response from an immune to a non-immune animal by the adoptive transfer of lymphoid cells.
Ref. 16 - pp. 331-332

425. The theory on which hyposensitization is based is to increase the antigenic doses at varying intervals in an attempt to develop blocking (γG) antibody protection against the antigen. Ref. 16 - p. 367

426. The autoimmune diseases are groups of disorders which have in common the manifestation of antibody, but not delayed hypersensitivity, to body constituents (autoimmune phenomena).
Ref. 16 - p. 430

427. Vaccines shown to be effective in localized viral infections are those which stimulate the local γA responses.
Ref. 16 - p. 489

428. Live vaccines contain antigens which are capable of sustained persistence and prolonged immunogenic effect. Ref. 16 - p. 490

429. Antibody molecules (IgG) are bivalent and monospecific; on the other hand, natural antigen molecules (proteins) are multivalent and commonly multispecific. Ref. 13 - p. 8

430. The over-all function of the thymus in immunity appears to be that of regulating maturation of stem cells so as to maintain the normal pool of immunocompetent cells. Ref. 13 - p. 45

431. Germinal centers in lymph nodes are collections of proliferating cells that result as a consequence of antibody production.
Ref. 13 - p. 54

432. Single antibody-forming cell can be detected by the Jerne (Jerne-Nordin) technique but not by the rosette (immuno-cyto-adherence) test.
Ref. 13 - p. 59

433. According to the Clonal Selection Theory of antibody formation, a single cell can form antibodies of but one specificity, i.e., against but one antigen-determinant or related determinants.
Ref. 13 - p. 65

434. The Jones-Mote sensitivity differs from a true delayed sensitivity in that the former is transitory and is commonly replaced by an Arthus type sensitivity, which coincides with the appearance of IgG antibodies in the serum. Ref. 13 - p. 202

435. Many of the intracellular pathogens produce chronic diseases in which effective immunity tends to be cellular, whereas infection with extracellular pathogens commonly leads to rapid development of humoral antibodies and specific immunity. Ref. 13 - p. 317

SECTION II - IMMUNOLOGY

MATCH EACH OF THE FIVE ITEMS LISTED BELOW WITH THE MOST APPROPRIATE STATEMENTS OR CHOICES. EACH ITEM MAY BE USED ONLY ONCE:

  A. Sheep red cells
  B. Incomplete antibody
  C. Cow serum
  D. Human O group RBC
  E. Precipitin reaction in gel

436. ___ Ouchterlony double diffusion
437. ___ Antiglobulin reaction
438. ___ Cold agglutinins
439. ___ Amboceptor
440. ___ Conglutinating complement adsorption test (CCAT)
　　　　　　　　　　　　　Ref. 2 - pp. 366, 369, 370, 379, 368

  A. Feeding tolerance
  B. Fetal tolerance
  C. Genetic tolerance
  D. Parity tolerance
  E. Split tolerance

441. ___ Failure to develop one but not the other of the following: delayed sensitivity and humoral antibody response
442. ___ Tolerance to foreign antigens acquired by their ingestion
443. ___ Tolerance of the mother to foreign antigens of the fetus as the result of pregnancy
444. ___ Tolerance acquired by exposure to antigens during fetal life
445. ___ Tolerance resulting from genetic identity
　　　　　　　　　　　　　Ref. 13 - p. 343

  A. Congenital
  B. TAB vaccine
  C. Inherent
  D. Flury vaccine
  E. 17D

446. ___ Salmonella vaccine
447. ___ Viable rabies vaccine
448. ___ Passively acquired resistance
449. ___ Native resistance
450. ___ Yellow fever vaccine　　　Ref. 7 - pp. 229, 231, 214, 214, 231

  A. Coombs' reaction
  B. "Lattice" theory
  C. Wasserman test
  D. Arthus reaction
  E. Urticaria

451. ___ Reaction between the lipoid antigen and the reagin in the serum
452. ___ Antibody to nonagglutinating antibody
453. ___ Due to precipitating antibody
454. ___ Aggregation is ascribed to specific attraction between antigen and antibody
455. ___ Due to either precipitating or nonprecipitating antibody in relatively small amounts
　　　　　　　　　　　　　Ref. 10 - pp. 406, 374, 416, 375, 415, 416

SECTION II - IMMUNOLOGY

   A. Allograft (homograft)
   B. Xenograft
   C. Autograft
   D. Syngraft (isograft)
   E. Heterotopic graft

456. ___ Graft derived from the same animal to which it is transplanted
457. ___ Graft derived from genetically dissimilar individuals of same species
458. ___ Graft derived from an animal of a different species from that of the recipient
459. ___ Graft placed in unnatural anatomic position
460. ___ Graft derived from genetically identical or near-identical animals

Ref. 13 - pp. 340, 339, 347, 210, 346

SECTION III - BACTERIOLOGY

FOR EACH OF THE FOLLOWING MULTIPLE CHOICE QUESTIONS CHOOSE THE ONE MOST APPROPRIATE ANSWER:

461. THE FOLLOWING ORGANISMS OBTAIN THEIR ENERGY FROM INORGANIC COMPOUNDS AND THEIR CARBON FROM CO_2:
   A. Chemo-autotrophs
   B. Photo-autotrophs
   C. Heterotrophs
   D. Chemoheterotrophs
   E. Photoheterotrophs
   Ref. 2 - p. 132

462. THOSE ORGANISMS WHICH USE LIGHT AS AN ENERGY SOURCE AND ORGANIC COMPOUNDS AS A CARBON SOURCE ARE KNOWN AS:
   A. Photo-autotrophs
   B. Photoheterotrophs
   C. Chemo-autotrophs
   D. Chemoheterotrophs
   E. Facultative autotrophs
   Ref. 2 - p. 132

463. A TYPICAL REPRESENTATIVE OF CHEMO-AUTOTROPHIC BACTERIA WHICH OXIDIZES NITRITE TO NITRATE:
   A. Nitrobacter
   B. Nitrosomonas
   C. Nitrosococcus
   D. Leptothrix
   E. Hydrogenomonas
   Ref. 2 - p. 134

464. THE FOLLOWING SPECIES OF BACTERIA IS COMMONLY ACTIVE IN THE NATURAL SOURING OF MILK:
   A. Bacterium (lactis) aerogenes
   B. Bacillus subtilis
   C. Proteus vulgaris
   D. Staphylococcus albus
   E. Staphylococcus citreus
   Ref. 2 - p. 323

465. THE ORGANISM MOST CLOSELY ALLIED TO CERTAIN AEROBES WHICH MAY BRING ABOUT THE OCCASIONAL FORMATION OF BUTYRIC ACID IN MILK:
   A. Bacterium aerogenes
   B. Escherichia coli
   C. Staphylococcus aureus
   D. Bacillus subtilis
   E. Streptococcus pyogenes
   Ref. 2 - p. 324

466. A RELATIVELY AVIRULENT STAPHYLOCOCCUS WHICH IS PART OF THE NORMAL FLORA OF THE HUMAN SKIN:
   A. Staphylococcus aerogenes
   B. Staphylococcus aureus
   C. Micrococcus tetragenus
   D. Staphylococcus citreus
   E. Staphylococcus albus
   Ref. 2 - p. 423

467. A SAPROPHYTIC MICROCOCCUS NAMED FOR ITS TENDENCY TO FORM CUBICAL PACKETS OF EIGHT CELLS:
   A. Micrococcus tetragenus
   B. Sarcina lutea
   C. Staphylococcus albus
   D. Staphylococcus citreus
   E. None of the above
   Ref. 2 - p. 426

468. A NON-PATHOGENIC STREPTOCOCCUS ABUNDANT IN NATURALLY SOURED MILK, PARTICULARLY WHEN THE ACIDITY IS HIGH:
   A. Streptococcus lacticus
   B. Streptococcus agalactiae
   C. Streptococcus salivarius
   D. Streptococcus equi
   E. Streptococcus bovis
   Ref. 2 - p. 323

469. THE BACTERIOLOGICAL GRADING OF MILK IS WIDELY BASED ON:
   A. Plate count
   B. Coliform count
   C. Total number of pathogenic bacteria
   D. Result of methylene blue reduction test
   E. Cell count and sediment test
   Ref. 2 - p. 325

SECTION III - BACTERIOLOGY

470. A NON-PATHOGENIC STAPHYLOCOCCUS PRODUCING A LEMON-YELLOW PIGMENT:
    A. Staphylococcus aerogenes
    B. Staphylococcus aureus
    C. Staphylococcus citreus
    D. Staphylococcus albus
    E. Micrococcus tetragenus
    Ref. 2 - p. 423

471. THE FOLLOWING SAPROPHYTIC MICROCOCCUS PRODUCES A BRIGHT YELLOW PIGMENT:
    A. Sarcina lutea
    B. Micrococcus tetragenus
    C. Staphylococcus aerogenes
    D. Staphylococcus aureus
    E. Staphylococcus albus
    Ref. 2 - p. 426

472. THE HOMOFERMENTATIVE GROUP OF LACTOBACILLI PRODUCES LARGE AMOUNTS OF:
    A. Propionic acid
    B. Carbon dioxide
    C. Acetic acid
    D. Ethanol
    E. Lactic acid
    Ref. 2 - p. 591

473. A HOMOFERMENTATIVE LACTOBACILLUS THOUGHT TO BE IDENTICAL WITH THE DÖDERLEIN BACILLUS:
    A. Lactobacillus bifidus
    B. Lactobacillus bulgaricus
    C. Lactobacillus brevis
    D. Lactobacillus acidophilus
    E. Lactobacillus buchneri
    Ref. 2 - p. 592

474. MOST BUTTER IS MADE FROM CHURNED CREAM SOURED BY THE FOLLOWING SPECIES SELECTED PRIMARILY FOR RAPID LACTIC ACID PRODUCTION:
    A. Leuconostoc citrovorum
    B. Lactobacillus brevis
    C. Bacterium linens
    D. Streptococcus lactis
    E. Saccharomyces cerevisiae
    Ref. 3 - p. 564

475. THE AROMA OF BUTTER IS PRODUCED FROM THE ACTION ON CITRIC ACID BY:
    A. Leuconostoc citrovorum
    B. Lactobacillus brevis
    C. Bacterium linens
    D. Streptococcus lactis
    E. Saccharomyces cerevisiae
    Ref. 3 - p. 564

476. PROPIONIBACTERIUM IS ACTIVE IN THE RIPENING AND FLAVORING OF:
    A. Cottage cheese
    B. Limburger cheese
    C. Gorgonzola cheese
    D. Camembert cheese
    E. Swiss cheese
    Ref. 3 - p. 566

477. ACETOBACTER PRODUCES VINEGAR BY OXIDIZING:
    A. Acetone
    B. Acetic acid
    C. Ethyl alcohol
    D. Propyl alcohol
    E. Butyric acid
    Ref. 3 - p. 588

478. NITRIFICATION REFERS TO THE:
    A. Oxidation of ammonia to nitrites, and nitrites to nitrates in the soil
    B. Fixation of atmospheric nitrogen into organic compounds
    C. Reduction of nitrates to nitrites and ammonia
    D. Hydolysis of urea to ammonia
    E. Synthesis of protein from amino acids
    Ref. 3 - p. 541

479. DIAPER RASH OCCURS AS A RESULT OF:
    A. Nitrification
    B. Fermentation
    C. Putrefaction
    D. Ammonification
    E. Denitrification
    Ref. 3 - p. 540

SECTION III - BACTERIOLOGY

480. SECRETIONS OF THE PREPUCE MAY CONTAIN:
   A. Streptococcus lactis
   B. Streptococcus thermophilus
   C. Mycobacterium smegmatis
   D. Mycobacterium phlei
   E. Mycobacterium stercoris
   Ref. 3 - p. 485

ANSWER THE FOLLOWING QUESTIONS BY USING THE KEY OUTLINED BELOW:
   A. If both statement and reason are true and related cause and effect
   B. If both statement and reason are true but not related cause and effect
   C. If the statement is true but the reason is false
   D. If the statement is false but the reason is true
   E. If both statement and reason are false

481. Döderlein's bacillus aids in the natural defenses against infection BECAUSE it contributes to the acidity of the vaginal secretions.
   Ref. 2 - p. 592

482. At one time it was supposed that a lactobacillus intestinal flora was preferable to a coliform proteolytic flora BECAUSE such a flora favors general health.
   Ref. 2 - p. 591

483. The acid-fast smegma bacillus may easily be distinguished morphologically from the tubercle bacillus BECAUSE smegma bacilli grow much more rapidly.
   Ref. 2 - p. 683

484. The sulfate-reducing bacteria are able to utilize sulfates as hydrogen acceptors BECAUSE they are found in polluted water.
   Ref. 3 - p. 544

485. A sample of milk that does not have phosphatase present cannot be regarded as safely pasteurized BECAUSE certain lactobacilli produce phosphatase unaffected by pasteurization.
   Ref. 3 - p. 560

486. The reductase test provides a rough approximation of the number and kinds of living bacteria present in raw milk BECAUSE actively growing bacteria maintain a high oxidation-reduction potential in the medium.
   Ref. 3 - p. 561

487. In making cottage cheese, starters containing mixtures of Leuconostoc citrovorum, L. dextranicum, S. lactis and the like are added to pasteurized milk BECAUSE rennet formed by the fermentation of lactose coagulates the casein.
   Ref. 3 - pp. 564-565

488. Spoilage of fresh meat has been prevented for days and even weeks by intravenous injections of antibiotics just before or after killing BECAUSE the antibiotic is distributed throughout the tissues by the blood stream.
   Ref. 3 - p. 570

489. Oysters for sale raw have greatly reduced bacterial count BECAUSE they are "floated" (allowed to remain for some hours or days) in clean chlorinated water.
   Ref. 3 - p. 572

490. Once thawed, frozen foods should be used immediately BECAUSE on thawing they may undergo rapid deterioration.
   Ref. 3 - p. 577

SECTION III - BACTERIOLOGY

ANSWER THE FOLLOWING QUESTIONS BY USING THE KEY OUTLINED BELOW:
A. If only A is correct
B. If only B is correct
C. If both A and B are correct
D. If neither A nor B is correct

491. MORPHOLOGICALLY, THE PHOTOSYNTHETIC BACTERIA ARE LIKE:
  A. Filamentous organisms
  B. Typical cocci or bacilli
  C. Both
  D. Neither
  Ref. 3 - p. 470

492. BACTERIAL PHOTOSYNTHESIS OCCURS:
  A. Only under anaerobic conditions
  B. In the presence of light
  C. Both
  D. Neither
  Ref. 3 - p. 470

493. THE FOLLOWING CONSTITUTE THE "NORMAL FLORA" OF MARKET MILK:
  A. Yeasts
  B. Pathogenic bacteria
  C. Both
  D. Neither
  Ref. 3 - p. 556

494. AN UNDESIRABLE CONDITION OF MILK KNOWN AS "RED MILK" IS CAUSED BY:
  A. Serratia marcescens
  B. Alcaligenes viscolactis
  C. Both
  D. Neither
  Ref. 3 - p. 556

495. "BLUE MILK" OFTEN RESULTS FROM THE PRESENCE OF:
  A. Pseudomonas syncyanea
  B. Klebsiella aerogenes
  C. Both
  D. Neither
  Ref. 3 - p. 556

ANSWER THE FOLLOWING QUESTIONS BY USING THE KEY OUTLINED BELOW:
A. If A is greater in frequency or magnitude than B
B. If B is greater in frequency or magnitude than A
C. If A and B are approximately equal

496. NUMBER OF MICROORGANISMS IN THE AIR IN:
  A. Humid weather
  B. Dry weather
  Ref. 3 - p. 554

497. THE PHOTOSYNTHETIC PROCESS IN:
  A. Green plants
  B. Photosynthetic bacteria
  Ref. 3 - p. 471

498. UTILIZATION OF VITAMIN B_{12} BY LACTOBACILLUS LACTIS:
  A. Aerobically
  B. Anaerobically
  Ref. 3 - p. 594

499. ACID PRODUCTION FROM GLUCOSE BY LACTOBACILLUS PLANTARUM IN MEDIUM:
  A. Containing nicotinic acid
  B. Lacking nicotinic acid
  Ref. 3 - pp. 592-593

500. EFFECTIVENESS OF PASTEURIZATION AT:
  A. 63° C for 30 minutes
  B. 71.6° to 80° C for 15 to 30 seconds
  Ref. 3 - pp. 556-558

SECTION III - BACTERIOLOGY

FOR EACH OF THE FOLLOWING MULTIPLE CHOICE QUESTIONS, CHOOSE THE ONE MOST APPROPRIATE ANSWER:

501. THE PRINCIPLE ENDOTOXIN FROM WHICH STREPTOCOCCI OBTAIN THEIR INVASIVENESS:
   A. Streptokinase
   B. Streptodornase
   C. Streptolysin
   D. Erythrogenic toxin
   E. None of the above
   Ref. 2 - p. 431

502. THE ORGANISM MOST COMMONLY CAUSING SUBACUTE BACTERIAL ENDOCARDITIS:
   A. Staphylococcus
   B. Pneumococcus
   C. Beta-hemolytic streptococcus
   D. Alpha-hemolytic streptococcus
   E. Gram-negative organisms resistant to penicillin
   Ref. 2 - p. 440

503. PNEUMOCOCCI CAN BEST BE DIFFERENTIATED FROM THE STREPTOCOCCI BY:
   A. Gram stain
   B. Type of hemolysis
   C. Clinical disease
   D. Growth characteristics
   E. Bile solubility
   Ref. 2 - p. 455

504. GONOCOCCUS IS:
   A. Demonstrated using dark field microscopy
   B. Easily cultured
   C. Gram-positive
   D. Nonmotile
   E. Never intracellular
   Ref. 2 - p. 463

505. DIFFERENTIATION OF THE ENTERIC BACILLI IS BASED ON:
   A. Colony appearance
   B. Gram stain
   C. Microscopic appearance
   D. Biochemical reactions
   E. Growth rate
   Ref. 2 - p. 480

506. A USEFUL PRIMARY DIFFERENTIATION CAN BE MADE BETWEEN PATHOGENIC AND NON-PATHOGENIC ENTERIC BACTERIA ON THE BASIS OF:
   A. Lactose fermentation
   B. Glucose fermentation
   C. H_2S production
   D. Gelatin liquefaction
   E. Indol formation
   Ref. 2 - p. 483

507. THE MOST COMMON SITE OF INFECTION WITH THE COLIFORM BACILLI IS THE:
   A. Upper respiratory tract
   B. Lower respiratory tract
   C. Urinary tract
   D. Gallbladder
   E. Colon
   Ref. 2 - p. 487

508. THE MAJOR DIFFERENCE BETWEEN PNEUMOCOCCAL PNEUMONIA AND PNEUMONIA DUE TO FRIEDLANDER'S BACILLUS IS:
   A. The greater incidence of Friedlander's bacillus pneumonia
   B. Low case fatality rate of pneumonia due to Friedlander's bacillus
   C. The greater tendency to cause necrotic lesions in Friedlander's pneumonia
   E. Absence of lung damage in Friedlander's pneumonia
   Ref. 2 - p. 488

SECTION III - BACTERIOLOGY

509. WHICH OF THE FOLLOWING STATEMENTS CONCERNING SALMONELLA ANTIGENS IS INCORRECT?:
A. "H" antigens are inactivated by heating over 60° C
B. Antibodies to "H" antigens are predominantly 7S
C. "O" antigens are resistant to prolonged heating at 100° C
D. Antibodies to "O" antigens are predominantly 19S
E. "O" antigens are prepared from only motile bacilli
Ref. 15 - p. 200

510. PHAGE TYPING OF Vi-CONTAINING SALMONELLA TYPHI IS USEFUL MAINLY AS:
A. A method of killing Salmonella, and thereby stopping epidemics in institutions
B. A therapeutic agent when chloromycetin resistance is present
C. Substitute for serological tests
D. An epidemiological tool
E. A public health measure to prevent typhoid fever in developing countries
Ref. 2 - p. 500

511. ANIMAL RESERVOIR FOR SALMONELLA TYPHIMURIUM:
A. Mice
B. Pigs
C. Rats
D. Horses
E. Reptiles
Ref. 2 - p. 501

512. ON SHIGELLA-SALMONELLA AGAR SALMONELLA APPEARS:
A. Dark because of glucose fermentation
B. Colorless because of hemolysis
C. Colorless because of no lactose fermentation
D. Dark because of no lactose fermentation
E. In S phase
Ref. 2 - p. 502

513. SALMONELLA INFECTION IS TRANSMITTED BY:
A. Ingestion
B. Inhalation
C. Contact
D. Droplet infection
E. None of the above
Ref. 2 - p. 503

514. THE AVERAGE CASE OF GASTROENTERITIS OF SALMONELLA ORIGIN IS ASSOCIATED WITH:
A. Short incubation, vomiting and diarrhea
B. A prolonged course with septicemia
C. A long incubation period with diarrhea, but no vomiting
D. A disease resembling typhoid fever
E. True intoxication
Ref. 2 - p. 504

515. TYPHOID ORGANISM CONTAINS THREE IMPORTANT ANTIGENS: H, O, AND Vi. THE FOLLOWING ARE BELIEVED TO HAVE GREATEST IMMUNOGENIC PROPERTY:
A. All three
B. H and O
C. O and Vi
D. Vi and H
E. Only one of the antigens
Ref. 2 - p. 512

516. WHEN IMMUNIZED AGAINST TYPHOID FEVER, ONE DEVELOPS:
A. Passive immunity
B. Hypersensitivity
C. Absolute immunity
D. A severe reaction
E. None of the above
Ref. 2 - p. 512

517. WHEN CULTURED ON A SOLID MEDIUM, MYCOBACTERIUM TUBERCULOSIS APPEARS:
A. Motile
B. Smooth
C. Metallic
D. Dry and granular
E. Moist
Ref. 2 - p. 664

518. THE CHOLERA VIBRIO:
   A. Is strongly anaerobic
   B. Grows best at 25° C
   C. Has a marked tolerance for alkaline pH
   D. Grows over a wide range but best at a slight acid pH
   E. Is nutritionally fastidious   Ref. 2 - p. 531

519. CHOLERA HAS HISTORICALLY BEEN ENDEMIC IN:
   A. China   D. France
   B. India   E. Mexico
   C. Germany   Ref. 2 - p. 530

520. THE SHORTEST DOUBLING TIME OBSERVED FOR THE TUBERCLE BACILLUS IN RICH MEDIA IS APPROXIMATELY:
   A. 20 minutes   D. 6 hours
   B. 1 hour   E. 12 hours
   C. 3 hours   Ref. 12 - p. 844

521. BRUCELLA ORGANISMS ARE:
   A. Gram-positive rods
   B. Gram-negative rods
   C. Gram-positive cocco-bacillary forms
   D. Gram-negative cocco-bacillary forms
   E. Acid-fast as well as gram-negative
                       Ref. 2 - p. 547

522. BRUCELLOSIS IS MOST COMMONLY TRANSMITTED IN THE U.S.A. BY:
   A. Flies   D. Raw milk
   B. Man to man contact E. None of the above
   C. Water   Ref. 2 - p. 552

523. MICROSCOPIC MORPHOLOGY OF PASTEURELLA PESTIS REVEALS:
   A. Colonies with a drop-like appearance
   B. Gram-negative colonies
   C. Gram-negative bacilli with a tendency toward bipolar staining
   D. Gram-negative bacilli with a tendency toward spore formation
   E. Intermediate Ziehl-Nielson staining properties
                       Ref. 2 - pp. 559-560

524. SYLVATIC PLAGUE REFERS TO A(N):
   A. Biblical epidemic among the Egyptians
   B. Venereal disease epidemic traced back to sylvatic times
   C. Epidemic among animals with Pasteurella tularensis
   D. Epidemic among animals with Pasteurella pestis
   E. Epidemic among animals with glanders bacillus
                       Ref. 2 - p. 563

525. PLAGUE HAS ITS RESERVOIR IN RATS, BUT IS TRANSMITTED BY:
   A. Ticks   D. Fleas
   B. Mosquitoes   E. Lice
   C. Mites   Ref. 2 - p. 563

526. FOOD POISONING CAUSED BY STAPHYLOCOCCUS AUREUS IS DUE TO PRODUCTION OF:
   A. Hemolysin   D. Leukocidin
   B. Lethal toxin   E. Exotoxin
   C. Enterotoxin   Ref. 6 - p. 395

SECTION III - BACTERIOLOGY

527. THE MOST DANGEROUS TYPES OF INFECTIONS CAUSED BY STAPHYLOCOCCI ARE:
    A. Cystitis and pyelonephritis
    B. Venereal disease and vaginitis
    C. Impetigo contagiosa and carbuncles
    D. Meningitis and pneumonia
    E. Bronchitis and bronchiectasis   Ref. 6 - p. 398

528. SCARLET FEVER IS CAUSED BY:
    A. Staphylococci
    B. Streptococci
    C. Coxsackie virus, Group B
    D. Adenovirus
    E. Pneumococci
    Ref. 6 - p. 377

529. THE MOST DANGEROUS FORM OF PUERPERAL SEPSIS IS CAUSED BY:
    A. Staphylococcus aureus
    B. Clostridium tetani
    C. E. coli
    D. Gonococci
    E. Group A hemolytic streptococci   Ref. 6 - p. 384

530. THE MOST RELIABLE TEST FOR DIFFERENTIATING PNEUMOCOCCI FROM OTHER COCCAL FORMS:
    A. Gram stain
    B. Autolysis in bile
    C. Hanging drop preparation
    D. Hemolysis
    E. Fermentation of glucose   Ref. 6 - p. 364

531. PNEUMOCOCCI CAN BE TYPED BY THE:
    A. Swelling of the capsule in the presence of type-specific antiserum
    B. Precipitation of the specific capsular polysaccharides
    C. Bile solubility of the organism
    D. Agglutination of the intact organism
    E. Transformation of specific types   Ref. 6 - p. 366

532. THE ESSENTIAL ANTIGEN IN THE PNEUMOCOCCUS WHICH DETERMINES BOTH ITS VIRULENCE AND SPECIFIC TYPE IS THE:
    A. Somatic carbohydrate
    B. Nucleoprotein structure
    C. Flagellar carbohydrate
    D. Thermolabile leukocidin
    E. Capsular polysaccharide
    Ref. 6 - p. 366

533. THE PNEUMOCOCCUS:
    A. Usually occurs in chains
    B. Is a spore-forming organism
    C. Is motile with one terminal flagellum
    D. Capsules are produced by virulent forms
    E. Grows best in a markedly acid pH   Ref. 6 - p. 364

534. ENCAPSULATED COLONIES OF HAEMOPHILUS INFLUENZAE:
    A. Ferment glucose with the production of acid and gas
    B. Grow profusely on blood agar plates
    C. Are designated as S
    D. Ferment lactose
    E. Cause Asian flu   Ref. 6 - p. 417

SECTION III - BACTERIOLOGY

535. MENINGITIS DUE TO H. INFLUENZAE OCCURS MOST FREQUENTLY IN:
    A. Infants under 2 months
    B. Children
    C. Adolescents
    D. Adults
    E. Patients with debilitating conditions
    Ref. 6 - p. 418

536. HAEMOPHILUS DUCREYI IS THE CAUSATIVE AGENT OF:
    A. Hard chancre
    B. Urethritis
    C. Tropical bubo
    D. Soft chancre
    E. Granuloma inguinale
    Ref. 6 - p. 420

537. BORDETELLA PERTUSSIS:
    A. Grows in and on the mucous membranes of the respiratory tract
    B. Is resistant to drying
    C. Is anaerobic
    D. Requires X and V factors for growth
    E. Is susceptible to the action of sulfonamides and penicillin
    Ref. 6 - p. 425

538. PHASE I B. PERTUSSIS IS PRESENT IN GREATEST ABUNDANCE DURING:
    A. Spasmodic stage
    B. Catarrhal stage
    C. Convalescent stage
    D. All of the above
    E. None of the above
    Ref. 6 - p. 425

539. CHARACTERISTICALLY, NEISSERIA MENINGITIDIS IS A:
    A. Gram-positive intracellular diplococcus
    B. Gram-negative intracellular diplococcus
    C. Facultative anaerobe
    D. Gram-positive extracellular diplococcus
    E. Gram-negative extracellular diplococcus
    Ref. 6 - p. 404

540. THE MAJORITY OF GONOCOCCAL INFECTIONS OCCURS IN THE GENITAL TRACTS BUT THE ORGANISMS HAVE ALSO GIVEN RISE TO:
    A. Septicemia
    B. Arthritis
    C. Osteomyelitis
    D. Endocarditis
    E. All of the above
    Ref. 6 - p. 411

541. OPHTHALMIA NEONATORUM:
    A. It is caused exclusively by N. gonorrheae
    B. If untreated it may lead to blindness
    C. It may transmit gonorrhea to the mother during delivery
    D. It may be transmitted to siblings
    E. It frequently causes strabismus
    Ref. 6 - p. 412

542. CORYNEBACTERIUM DIPHTHERIAE:
    A. Is gram-negative
    B. Grows moderately well under anaerobic conditions
    C. Is resistant to penicillin
    D. Forms spores
    E. Exhibits marked pleomorphism
    Ref. 6 - p. 430

543. MYCOBACTERIUM TUBERCULOSIS HAS ALL THE FOLLOWING CHARACTERISTICS, EXCEPT:
    A. Non-motile
    B. Difficult to stain
    C. Easily decolorized by mineral acids
    D. No capsules
    E. Non-spore forming
    Ref. 6 - p. 454

SECTION III - BACTERIOLOGY

544. IN MAN, THE MOST COMMON METHOD OF ENTRY OF THE TUBERCLE BACILLUS IS:
     A. Inhalation
     B. Ingestion
     C. Directly through the skin
     D. Hematogenous
     E. Renal
     Ref. 6 - p. 462

545. REINFECTION TUBERCULOSIS IS ALMOST EXCLUSIVELY A DISEASE OF THE:
     A. Lungs
     B. Bones
     C. Joints
     D. Brain
     E. All of the above
     Ref. 6 - p. 463

546. THE PROGNOSIS IS POOR IN THE FOLLOWING TYPE OF LEPROSY:
     A. Lepromatous
     B. Tuberculoid
     C. Diomorphous
     D. Indeterminate
     E. All of the above
     Ref. 6 - p. 448

547. SALMONELLA TYPHI:
     A. Produces acid with gas from glucose
     B. Produces acid with gas from mannitol
     C. H_2S production is variable
     D. Ferments lactose
     E. Is non-motile
     Ref. 6 - p. 510

548. THE HIGHEST INCIDENCE OF CHRONIC CARRIERS OF TYPHOID IS FOUND AMONG:
     A. Children
     B. Young adults
     C. Cooks and foodhandlers
     D. Middle-aged men
     E. Middle-aged women
     Ref. 6 - p. 513

549. IT IS POSSIBLE TO DIFFERENTIATE SALMONELLA FROM SHIGELLA BY THE FOLLOWING PROPERTY:
     A. Gram stain
     B. Motility
     C. Presence of a capsule
     D. One is aerobic
     E. Shape of spore forms
     Ref. 6 - p. 504

550. THE ANTIGENIC STRUCTURE OF SHIGELLA:
     A. Is much more complex than Salmonella
     B. Contains numerous major somatic antigens
     C. Is less complex than Salmonella
     D. The minor antigens are specific
     E. The minor O antigens never cross-react with the O antigens of other enteric bacilli
     Ref. 6 - p. 524

551. THE FOLLOWING STATEMENTS ARE CORRECT FOR SERRATIA MARCESCENS, EXCEPT:
     A. It was known in the older literature as Bacillus prodigiosus
     B. It was considered as a harmless saprophyte until recently
     C. It is being isolated in the laboratory with increasing frequency
     D. All the strains produce a red pigment
     E. Various clinical conditions have been associated with the organism
     Ref. 6 - p. 534

552. E. COLI IS CHARACTERIZED BY:
     A. Fermentation of carbohydrates to lactic acid, CO_2, and hydrogen
     B. Fermentation of carbohydrates to citric acid, carbon monoxide and water
     C. Non-fermentation of glucose and lactose
     D. Production of pink colonies on Endo's medium
     E. Production of alpha hemolysis on blood agar
     Ref. 6 - p. 538

SECTION III - BACTERIOLOGY

553. ONE OF THE FOLLOWING IS NOT CHARACTERISTIC OF FRANCISELLA TULARENSIS:
     A. Pleomorphism
     B. Gram-negative
     C. Aerobic environment
     D. Many antigenic types
     E. Resistant to sulfonamides and penicillin
     Ref. 6 - p. 567

554. ALL OF THE FOLLOWING STATEMENTS DESCRIBE PLAGUE, EXCEPT:
     A. The etiologic agent is now known as Yersinia pestis
     B. Man himself has been shown to be a healthy carrier recently in South Vietnam
     C. There are three clinical types
     D. The pneumonic form is also transmitted by the bite of the infected fleas
     E. Formalin-killed Y. pestis vaccine has been successful in South Vietnam
     Ref. 6 - p. 577

555. BACILLUS ANTHRACIS:
     A. Causes anthracosis which can be prevented by using electricity instead of coal
     B. Causes "wool-sorter's disease" and is spread by contaminated wool, horse hair and hides
     C. Forms spores which are extremely heat sensitive
     D. Is a small straight rod which grows under anaerobic conditions
     E. Causes a mild enteritis and development of a carrier state
     Ref. 6 - p. 585

556. PNEUMOCOCCAL CAPSULES TEND TO BE LARGEST DURING:
     A. Lag phase
     B. Exponential phase
     C. Stationary phase
     D. All phases
     E. After death
     Ref. 12 - p. 685

557. THE SEROLOGICAL TYPES OF PNEUMOCOCCI THAT HAVE BEEN DIFFERENTIATED BY THE IMMUNOLOGICALLY DISTINCT POLYSACCHARIDES IN THE CAPSULES NUMBER:
     A. 25
     B. 35
     C. 50
     D. 65
     E. 82
     Ref. 6 - p. 366

558. SPUTUM SMEARS FROM PNEUMONIA PATIENTS SHOULD BE STAINED WITH GRAM STAIN TO DISTINGUISH D. PNEUMONIAE FROM:
     A. Staphylococcus aureus
     B. Streptococcus pyogenes
     C. Klebsiella pneumoniae
     D. Hemophilus influenzae
     E. Neisseria meningitidis
     Ref. 12 - p. 698

559. KLEBSIELLA PNEUMONIAE:
     A. Is motile
     B. Is susceptible to penicillin
     C. Causes about 3 per cent of all acute bacterial pneumonias
     D. Is not found in healthy subjects
     E. Is morphologically distinguishable from A. aerogenes
     Ref. 12 - p. 772

560. PROLONGED SALMONELLA SEPTICEMIA IS MOST COMMONLY CAUSED BY:
     A. S. schottmülleri
     B. S. typhimurium
     C. S. choleraesuis
     D. S. enteritidis
     E. S. newport
     Ref. 12 - p. 777

SECTION III - BACTERIOLOGY

561. THE MOST COMMON CAUSE OF ENTERIC FEVER IN THE UNITED STATES IS:
    A. S. typhosa
    B. S. schottmülleri
    C. S. paratyphi
    D. S. typhimurium
    E. S. enteritidis
    Ref. 12 - p. 777

562. THE FOLLOWING FORM OF TULAREMIA RESULTS FROM PRIMARY INFECTION OF THE SKIN:
    A. Oculoglandular
    B. Ulceroglandular
    C. Pneumonic
    D. Typhoidal
    E. None of the above
    Ref. 12 - p. 807

563. IN NATURALLY ACQUIRED BRUCELLOSIS, THE ORGANISMS GAIN ENTRANCE TO THE BODY VIA THE:
    A. Alimentary tract
    B. Conjunctivae
    C. Broken skin
    D. All of the above
    E. None of the above
    Ref. 12 - p. 814

564. THE MOST COMMON CAUSE OF GAS GANGRENE IN CIVILIAN PRACTICE IS INFECTION OF THE UTERUS BY:
    A. C. novyi
    B. C. septicum
    C. C. histolyticum
    D. C. perfringens
    E. C. tertium
    Ref. 12 - p. 836

565. DONOVANIA GRANULOMATIS PRODUCES LARGE CAPSULES THAT CROSS-REACT IMMUNOLOGICALLY WITH THOSE OF:
    A. Klebsiella pneumoniae
    B. Diplococcus pneumoniae
    C. Hemophilus ducreyi
    D. All of the above
    E. None of the above
    Ref. 12 - p. 918

ANSWER THE FOLLOWING QUESTIONS BY USING THE KEY OUTLINED BELOW:
A. If A is greater in frequency and/or magnitude than B
B. If B is greater in frequency and/or magnitude than A
C. If A and B are approximately equal

566. THE SENSITIVITY TO PENICILLIN:
    A. Staphylococcus aureus
    B. Pneumococcus
    Ref. 6 - pp. 368, 399

567. CAUSE OF FOOD POISONING IN THE U.S.A.:
    A. Staphylococcus aureus
    B. Salmonella
    Ref. 6 - p. 395

568. IN THE PRESENCE OF AN ERYTHEMATOUS RASH, THE DIAGNOSIS OF SCARLET FEVER CAN BE CONFIRMED MORE FREQUENTLY BY THE USE OF:
    A. Dick test
    B. Schultz-Charlton test
    Ref. 6 - p. 380

569. RHEUMATIC FEVER MAY FOLLOW INFECTION WITH:
    A. Beta-hemolytic streptococcus, Group A, Type 12
    B. Beta-hemolytic streptococcus, Group A, assorted types
    Ref. 6 - p. 380

570. MOST CASES OF ACUTE HEMORRHAGIC GLOMERULONEPHRITIS FOLLOW INFECTION WITH:
    A. Beta-hemolytic streptococcus, Group A, Type 12
    B. Beta-hemolytic streptococcus, Group A, assorted types
    Ref. 6 - p. 381

SECTION III - BACTERIOLOGY

571. SUBACUTE BACTERIAL ENDOCARDITIS CAUSED BY:
　A. Streptococcus viridans
　B. Anaerobic streptococcus　　　Ref. 6 - p. 384

572. PHAGE GROUP STRAINS THAT PREDOMINATE IN HOSPITAL STAPHYLOCOCCAL INFECTIONS:
　A. Phage group II strains, e.g., phage 71
　B. Phage group I strains, e.g., phage 81
　　　　　　　　　　　　　　　Ref. 6 - p. 391

573. PRODUCTION OF ANTIBODIES AGAINST:
　A. Streptolysin-O
　B. Streptolysin-S　　　Ref. 6 - p. 375

574. THE COMPLEXITY OF THE ANTIGENIC STRUCTURE OF:
　A. Pneumococci
　B. Streptococci　　　Ref. 6 - p. 374

575. MOST CHARACTERISTIC OF PNEUMOCOCCAL INFECTIONS:
　A. Meningitis
　B. Lobar pneumonia　　　Ref. 6 - p. 368

576. THE REQUIREMENT OF X AND V FACTORS FOR GROWTH BY:
　A. Haemophilus influenzae
　B. Haemophilus aegyptius　　　Ref. 6 - p. 416

577. GROWTH OF H. INFLUENZAE ON:
　A. Blood agar
　B. Chocolate agar　　　Ref. 6 - p. 417

578. PASSIVE IMMUNITY TO H. INFLUENZAE INFECTIONS:
　A. One week-old child
　B. One year-old child　　　Ref. 6 - p. 418

579. TREATMENT OF CHOICE FOR H. INFLUENZAE MENINGITIS:
　A. Chlortetracycline
　B. Chloramphenicol　　　Ref. 6 - p. 418

580. MOTILITY OF:
　A. H. influenzae
　B. H. ducreyi　　　Ref. 6 - pp. 417, 420

581. RELIABILITY OF THE TUBERCULIN SKIN TEST OF:
　A. Tine
　B. Mantoux　　　Ref. 6 - p. 459

582. ACTIVE IMMUNIZATION AGAINST DIPHTHERIA WITH:
　A. Alum-precipitated toxoid
　B. Fluid toxoid　　　Ref. 6 - p. 435

583. SPECIFIC TREATMENT FOR DIPHTHERIA:
　A. Diphtheria antitoxin
　B. Penicillin　　　Ref. 6 - p. 435

584. RESISTANCE TO TUBERCULOUS DISEASE IN:
　A. Jews
　B. Non-Jews　　　Ref. 6 - p. 465

SECTION III - BACTERIOLOGY

585. SUSCEPTIBILITY OF MAN TO:
    A. Tuberculous infection
    B. Tuberculous disease   Ref. 6 - p. 462

586. SUSCEPTIBILITY TO DRUGS SUCH AS STREPTOMYCIN AND ISONIAZID:
    A. Mycobacterium kansasii
    B. Mycobacterium tuberculosis   Ref. 6 - p. 475

587. DRUG OF CHOICE FOR TYPHOID FEVER:
    A. Sulfonamides
    B. Chloramphenicol   Ref. 6 - p. 513

588. BASIC VIRULENCE OF SALMONELLA TYPHOSA:
    A. O antigen
    B. H antigen   Ref. 6 - p. 511

589. THE INCUBATION PERIOD OF:
    A. Salmonella food poisoning
    B. Staphylococcal food poisoning   Ref. 6 - pp. 395, 519

590. THE SEVERITY OF THE SALMONELLA INFECTIONS:
    A. Septicemic type
    B. Typhoidal type   Ref. 6 - p. 519

591. DIAGNOSIS OF SHIGELLA DYSENTERY BY:
    A. Positive blood cultures
    B. Positive stool cultures   Ref. 6 - p. 525

592. THE PREDOMINANT TYPE OF CLOSTRIDIUM BOTULINUM FOUND IN THE EASTERN UNITED STATES:
    A. Type B
    B. Type A   Ref. 6 - p. 606

593. THE MOST FREQUENT DISEASE CAUSED BY CLOSTRIDIUM PERFRINGENS:
    A. Food poisoning
    B. Gas gangrene   Ref. 6 - p. 595

594. MOST CASES AND DEATHS OF BOTULISM IN THE U.S.A.:
    A. Home-canned vegetables
    B. Smoked, salted, or spiced meats   Ref. 6 - p. 607

595. THE MOST USEFUL EXPERIMENTAL ANIMAL FOR INFECTION WITH TREPONEMA PALLIDUM:
    A. Rabbit
    B. Guinea pig   Ref. 6 - p. 645

ANSWER THE FOLLOWING QUESTIONS BY USING THE KEY OUTLINED BELOW:
A. If both statement and reason are true and related cause and effect
B. If both statement and reason are true but not related cause and effect
C. If the statement is true but the reason is false
D. If the statement is false but the reason is true
E. If both statement and reason are false

596. Very young colonies of S. aureus are colorless but it was given its name BECAUSE as growth takes place a golden yellow pigment is produced.
                                         Ref. 6 - p. 389

SECTION III - BACTERIOLOGY

597. The most common type of staphylococcal infection in man are pimples, boils, and carbuncles BECAUSE nearly 90% of staphylococci isolated from the skin as impetigo were susceptible to phage type 71.
Ref. 6 - p. 398

598. The spread of S. aureus can be decreased by gargling with antibiotic solutions BECAUSE the major method of spread of this organism is by droplet infection from the nasopharynx.   Ref. 6 - p. 400

599. The streptococci cause a greater variety of clinical types of disease than any other microorganism BECAUSE in addition to general infections, they cause specific diseases such as scarlet fever and erysipelas.
Ref. 6 - p. 377

600. The nasal carrier of streptococci is the dangerous carrier BECAUSE the ordinary throat carriers expel only a few streptococci into the air.
Ref. 6 - p. 377

601. Recurrences of acute hemorrhagic glomerulonephritis are rarely seen BECAUSE most of the patients die during the acute phase of the disease.
Ref. 6 - p. 381

602. Penicillin prophylaxis is not indicated in acute rheumatic fever patients BECAUSE it cannot prevent subsequent recurrences of the disease.
Ref. 6 - pp. 380-381

603. On blood plates after 24-48 hours' incubation, pneumococci can be confused with alpha-hemolytic streptococci BECAUSE both produce small round compact colonies surrounded by a zone of clear hemolysis.
Ref. 6 - p. 365

604. Pertussis organisms have been placed in a new genus Bordetella BECAUSE they do not require X and V factors for growth as the Hemophilus.
Ref. 6 - p. 424

605. Children should be actively immunized with pertussis vaccine about the third month of life BECAUSE children of this age do not produce antibodies as readily as children of nine months of age.
Ref. 6 - p. 426

606. Man is not infected with meningococci by way of the nasopharynx BECAUSE most of the deaths occur in cases of fulminating meningococcemia, with or without an accompanying meningitis.   Ref. 6 - pp. 405-406

607. Neisseria catarrhalis is an important pathogen BECAUSE it can produce an infection which duplicates the signs and symptoms of an infection with N. meningitidis.   Ref. 6 - p. 407

608. A definitive diagnosis of gonorrhea can be made from smears of pus from the male urethra BECAUSE the gonococci are found frequently in pairs and usually intracellularly in these smears.
Ref. 6 - p. 410

609. Development of a urethral discharge in the male three weeks after contact is not likely to be gonorrheal in origin BECAUSE the usual incubation period has been shown to be from three to five days.
Ref. 6 - p. 411

610. The Schick test has fallen into disrepute BECUASE false positive reactions are common in young children.   Ref. 6 - pp. 433-434

SECTION III - BACTERIOLOGY

611. Tubercle bacilli will not grow in the absence of oxygen BECAUSE they are obligate aerobes.　　　　　　　　　　　Ref. 6 - p. 455

612. Pasteurization temperatures are effective in eliminating tubercle bacilli from milk BECAUSE tubercle bacilli possess no greater resistance to moist heat than other bacteria.　　　Ref. 6 - p. 456

613. The prognosis is good in the lepromatous type of leprosy BECAUSE the lesions contain relatively few bacilli.　　Ref. 6 - pp. 448-449

614. Members of the Salmonella group are classified into species by their antigenic structures BECAUSE most Salmonella are motile.
　　　　　　　　　　　　　　　　　　　Ref. 6 - p. 518

615. Enrichment media are designed to suppress the normal flora of the stool and favor the multiplication of pathogens BECAUSE pathogens are often present in small numbers, e.g., late in disease and among carriers.
　　　　　　　　　　　　　　　　　　　Ref. 6 - p. 502

616. Determination of Vi antibody is useful in the detection of chronic typhoid carriers BECAUSE 75% of the carriers have Vi antibody in their serum.
　　　　　　　　　　　　　　　　　　　Ref. 6 - p. 511

617. Klebsiella pneumoniae can be grown easily in the laboratory BECAUSE it is aerobic and produces slimy, semifluid colonies.
　　　　　　　　　　　　　　　　　　　Ref. 6 - p. 530

618. E. coli is not thought to be pathogenic for man BECAUSE it is a normal inhabitant of the intestinal tract and occurs in greatest concentration in the region of the ileocecal valve.　　Ref. 6 - p. 538

619. Unlike typhoid fever, Asiatic cholera is not spread by contaminated water or food BECAUSE cholera cannot persist in an area where sanitation is adequate.　　　　　　　　　　　　　　Ref. 6 - pp. 553, 554

620. Clostridium tetani is an obligate anaerobe BECAUSE it loses its ability to produce toxin under aerobic conditions.
　　　　　　　　　　　　　　　　　　　Ref. 6 - p. 600

621. Haemophilus aegyptius is now known to be different from H. influenzae BECAUSE the former requires the X and V factors for growth.
　　　　　　　　　　　　　　　　　　　Ref. 6 - pp. 418-419

622. Diphtheroid bacilli are called as such BECAUSE of their morphologic resemblance to the diphtheria bacillus.　　Ref. 6 - p. 437

623. The tuberculin test becomes positive between the fourth and the twelfth week after infection BECAUSE it takes that length of time for the bacilli to spread through the lymphatics.　　　Ref. 6 - p. 463

624. BCG vaccine should be administered only to those individuals who are tuberculin-negative BECAUSE vaccination of tuberculin-positive individuals results in a superficial ulceration at the site of inoculation which persists for many weeks and injures the patient.
　　　　　　　　　　　　　　　　　　　Ref. 6 - p. 467

625. Most investigators believe that the classical type of infectious mononucleosis is caused by Listeria monocytogenes BECAUSE patients and experimental animals infected with L. monocytogenes always have the heterophile antibodies.　　　　　　Ref. 6 - p. 611

SECTION III - BACTERIOLOGY

626. Plague became known as the "black death" BECAUSE of its tendency to produce severe cyanosis in its terminal stages.
Ref. 12 - p. 802

627. The case fatality rate in untreated ulceroglandular tularemia is more than 30 percent BECAUSE the incubation period in tularemia ranges from 3 to 10 days. Ref. 12 - p. 807

628. Bacterial diseases due to H. influenzae are common in children BECAUSE the organism gains entrance to the tissues via the respiratory tract. Ref. 12 - p. 793

629. The brucella organisms exhibit marked pleomorphism BECAUSE smooth (S) variants produce demonstrable capsules and form small, translucent, glistening colonies. Ref. 12 - p. 812

630. The anthrax bacillus grows best aerobically BECAUSE aerobic conditions are required for sporulation. Ref. 12 - p. 820

631. Most species of Clostridium are obligate anaerobes BECAUSE all species of Clostridum form spores. Ref. 12 - p. 828

632. Although there is no evidence that the hemolysin, tetanolysin, plays a significant role in pathogenesis, the hemolysin is of immunological interest BECAUSE of its antigenic similarity to streptolysin O.
Ref. 12 - p. 834

633. Antibodies to somatic antigens of C. tetani are protective BECAUSE immunity to tetanus does not involve antitoxic antibody to the exotoxin.
Ref. 12 - p. 834

634. Listeria monocytogenes may be mistaken for a hemolytic streptococcus BECAUSE Listeria colonies grow well on blood agar and usually are surrounded by a narrow zone of B-hemolysis.
Ref. 12 - p. 916

635. Tuberculin-positive individuals should never be vaccinated BECAUSE they may have a severe reaction. Ref. 12 - p. 859

SELECT THE ITEM FROM COLUMN I WHICH IS UNRELATED TO THE FOUR OTHER ITEMS IN THIS COLUMN, CIRCLE THIS NUMBER. IDENTIFY THE ITEM IN COLUMN II TO WHICH THE FOUR RELATED ITEMS IN COLUMN I ARE ASSOCIATED. CIRCLE THIS LETTER.

I

II

636.
1. Syphilis
2. Gonorrhea
3. Wasserman reaction
4. Darkfield illumination
5. VDRL test

A. Treponema
B. Spirillum
C. Neisseria

Ref. 6 - pp. 642-644

637.
1. "Trench mouth"
2. Anaerobiosis
3. Exotoxin
4. Risus sardonicus
5. Lockjaw

A. Borrelia recurrentis
B. Borrelia vincentii
C. Clostridium tetani

Ref. 6 - pp. 600-601

SECTION III - BACTERIOLOGY

638.
1. Tellurite medium
2. Löffler's slant
3. Dick test
4. Exotoxin
5. β phage

A. Clostridium tetani
B. Corynebacterium diphtheriae
C. Hemolytic streptococcus

Ref. 6 - pp. 430-431

639.
1. Hemolytic
2. Enterotoxin
3. Food poisoning
4. Coagulase positive
5. Gram-negative

A. Staphylococcus aureus
B. Streptococcus viridans
C. Neisseria gonorrheae

Ref. 6 - pp. 392-395

640.
1. Scarlet fever
2. Epidemic sore throat
3. Rheumatic fever
4. Lobar pneumonia
5. Puerperal sepsis

A. Group A streptococcus
B. Pneumococcus
C. Staphylococcus aureus

Ref. 6 - pp. 380-384

641.
1. Transformation
2. Quellung method
3. Tetrads
4. Alpha type hemolysis
5. Ferments inulin

A. Streptococcus viridans
B. Pneumococcus
C. Staphylococcus aureus

Ref. 6 - pp. 365-366

642.
1. 2-10 percent CO_2
2. Chancre
3. Thayer-Martin medium
4. Oxidase test
5. 2 percent silver nitrate

A. Gonococcus
B. Treponema pallidum
C. Pneumococcus

Ref. 6 - pp. 410-412

643.
1. Stormy fermentation in milk
2. Glucose fermentation
3. Maltose fermentation
4. No change in milk
5. Gelatin liquefaction

A. Clostridium histolyticum
B. Clostridium perfringens
C. Clostridium tetani

Ref. 6 - p. 591

644.
1. Bubonic plague
2. Black death
3. Pneumonic plague
4. Buboes
5. Endocarditis

A. Yersinia
B. Brucella
C. Bordetella

Ref. 6 - p. 577

645.
1. Non-lactose fermenter
2. O and K antigens
3. Decompose urea
4. Non-motile
5. "Swarming"

A. Shigella
B. Salmonella
C. Proteus

Ref. 6 - p. 541

646.
1. H_2S production
2. Inhibition by dyes
3. Pleomorphic
4. Gram-negative
5. Coccobacillus

A. Yersinia
B. Brucella
C. Bordetella

Ref. 6 - p. 558

647.
1. Chancroid
2. Nonsporogenous
3. Motile
4. X and V factors
5. Contagious conjunctivitis

A. Brucella
B. Treponema
C. Hemophilus

Ref. 6 - pp. 416-420

SECTION III - BACTERIOLOGY

648.
1. Acid-fast stain
2. Much's granules
3. Spore forming
4. PPD
5. OT

A. Tuberculosis
B. Leprosy
C. Rat leprosy

Ref. 6 - pp. 454-458

649.
1. Mallein test
2. Spores
3. Glanders
4. Nonflagellated
5. Nonmotile

A. Actinobacillus mallei
B. Francisella tularensis
C. Bacillus anthracis

Ref. 6 - pp. 440-441

650.
1. Soft tick
2. Body louse
3. Vincent's angina
4. Relapsing fever
5. Chancre

A. Rickettsia
B. Treponema
C. Borrelia

Ref. 6 - pp. 652-654

651.
1. Icterus
2. Jaundice
3. Leptospira
4. Skin hemorrhages
5. Spirillum

A. Rat-bite fever
B. Weil's disease
C. Syphilis

Ref. 6 - p. 663

652.
1. El Tor
2. Rats
3. Diarrhea
4. "Rice water"
5. Contaminated water

A. Vibrio comma
B. Yersinia pestis
C. Salmonella typhosa

Ref. 6 - pp. 550, 553

653.
1. Gram-positive rod
2. Anaerobic
3. Spore forming
4. Pustule
5. Tanners of hide

A. Listeria monocytogenes
B. Bacillus anthracis
C. Clostridium perfringens

Ref. 6 - pp. 582, 584

654.
1. Ectoplasm
2. Cork-screw-like
3. Motile
4. Axial filaments
5. Spiral

A. Spirochetes
B. Chlamydia
C. Protozoa

Ref. 6 - p. 638

655.
1. Produce penicillinase
2. Produce gelatinase
3. Produce coagulase
4. Produce phosphatase
5. Produce H_2S

A. Staphylococcus aureus
B. Salmonella typhosa
C. Escherichia coli

Ref. 6 - pp. 392-393

656.
1. Erythrogenic toxin
2. Streptokinase
3. Hyaluronidase
4. Streptolysin-O
5. Enterotoxin

A. Staphylococcal metabolites
B. Streptococcal metabolites
C. Pneumococcal metabolites

Ref. 6 - p. 375

SECTION III - BACTERIOLOGY   77

657.
1. Clostridium botulinum
2. Clostridium septicum
3. Clostridium novyi
4. Clostridium perfringens
5. Clostridium histolyticum

A. Insignificant local infection
B. Food poisoning
C. Gas gangrene

Ref. 6 - p. 590

658.
1. Open ulcer
2. Rabbit reservoir
3. Dog reservoir
4. Hard tick
5. Deer fly

A. Tularemia
B. Bubonic plague
C. Rabies

Ref. 6 - pp. 569-570

659.
1. CO_2 requirement -
2. Urea+
3. Basic fuchsin (1:200)+
4. Crystal violet (1:400)+
5. Thionin (1:800)-

A. Brucella abortus
B. Brucella melitensis
C. Brucella suis

Ref. 6 - p. 559

660.
1. Wasserman test
2. Kolmer test
3. Kahn test
4. TPI test
5. Eagle test

A. Utilize Wasserman antigen
B. Utilize Reiter protein antigen
C. Utilize immobilizing antigen

Ref. 12 - p. 885

661.
1. Escherichia coli
2. Salmonellae
3. Klebsiella pneumoniae
4. Shigellae
5. Vibrio cholerae

A. Coliform bacilli
B. Members of Enterobacteriaceae
C. Nonpathogens

Ref. 12 - p. 756

662.
1. Proteus vulgaris
2. Proteus mirabilis
3. Shigella dysenteriae
4. Salmonella typhosa
5. Salmonella para B

A. Produce H_2S
B. Citrate utilization
C. Produce urease

Ref. 12 - p. 757

663.
1. S. hirschfeldii
2. S. choleraesuis
3. S. montevideo
4. S. typhosa
5. S. oranienburg

A. Kauffmann-White antigen group C_1
B. Kauffmann-White antigen group B
C. Kauffmann-White antigen group D

Ref. 12 - p. 763

664.
1. H. influenzae
2. H. parainfluenzae
3. H. ducreyi
4. H. suis
5. H. hemolyticus

A. Require X factor
B. Require V factor
C. Produce hemolysis

Ref. 12 - p. 790

665.
1. C. botulinum
2. C. tetani
3. C. perfringens
4. C. novyi
5. C. septicum

A. Ferment lactose
B. Liquefy gelatin
C. Ferment glucose

Ref. 12 - p. 830

SECTION III - BACTERIOLOGY

MATCH EACH OF THE LETTERED ITEMS WITH THE MOST APPROPRIATE STATEMENTS OR CHOICES. EACH ITEM MAY BE USED ONLY ONCE:

A. Pfeiffer's bacillus
B. Frankel's bacillus
C. Pyocyanea
D. Klebsiella pneumoniae
E. Bacille Calmette-Guérin

666. ___ Hemophilus influenzae
667. ___ Pseudomonas aeruginosa
668. ___ Mycobacterium
669. ___ Clostridium welchii
670. ___ Friedlander's bacillus   Ref. 2 - pp. 578, 589, 667, 633, 487

A. "Black death"
B. Pneumonia
C. Dysentery
D. Pustules
E. Scarlet fever

671. ___ Streptococcus pyogenes
672. ___ Diplococcus pneumoniae
673. ___ Pasteurella pestis
674. ___ Staphylococcus aureus
675. ___ Shigella   Ref. 2 - pp. 445, 453, 558, 424, 516

A. Droplet infection
B. Gastroenteritis
C. Intoxication
D. Colon bacillus
E. Diphtheroid

676. ___ Clostridium botulinum
677. ___ Salmonella typhimurium
678. ___ Corynebacterium hofmannii
679. ___ Escherichia coli
680. ___ Mycobacterium tuberculosis   Ref. 2 - pp. 645, 504, 660, 485, 668

A. Mantoux skin test
B. "Lockjaw"
C. Subacute bacterial endocarditis
D. Diphtheria bacillus
E. Undulant fever

681. ___ Gram-negative coccobacillus
682. ___ Gram-positive coccus
683. ___ Acid-fast bacillus
684. ___ Gram-positive rod with metachromatic granules
685. ___ Gram-positive sporulating rods   Ref. 2 - pp. 547, 449, 663, 674, 649, 625

SECTION III - BACTERIOLOGY

A. Schick test
B. Alum-precipitated toxoid
C. Dick test
D. Involution forms
E. Inaba and Ogawa serotypes

686. ___ Tetanus
687. ___ Diphtheria
688. ___ Plague
689. ___ Cholera
690. ___ Streptococcus   Ref. 2 - pp. 630, 657, 559, 541, 446

A. C-réactive protein (species-specific)
B. C carbohydrates (group specific) in cell wall
C. A polysaccharide in cell wall
D. B polysaccharide in cell wall
E. Fraction 1 in envelope or capsule

691. ___ β-hemolytic streptococci
692. ___ Staphylococcus albus
693. ___ Pneumococcus
694. ___ Pasteurella pestis
695. ___ Staphylococcus aureus   Ref. 12 - pp. 709, 731, 690, 803, 731

A. Requirement for cystine
B. "Chocolate agar"
C. Transmitted by bite of fleas
D. Bordet-Gengou plate
E. Enterotoxin

696. ___ P. pestis
697. ___ S. aureus
698. ___ P. tularensis
699. ___ B. pertussis
700. ___ N. gonorrhoeae   Ref. 12 - pp. 803, 730, 806, 798, 743

A. Crepitation on physical examination
B. Malignant pustule
C. Toxemia
D. "Millet seed"
E. Malta fever

701. ___ Brucellosis
702. ___ Tuberculosis
703. ___ Gas gangrene
704. ___ Anthrax
705. ___ Staphylococcus   Ref. 12 - pp. 812, 850, 836, 823, 737

SECTION III - BACTERIOLOGY

A. "Medusa head" colonies
B. Koch-Weeks bacillus
C. Pfeiffer phenomenon
D. Babes-Ernst bodies
E. Quellung reaction

706. ___ H. aegypticus
707. ___ D. pneumoniae
708. ___ B. anthracis
709. ___ C. diphtheriae
710. ___ V. cholerae   Ref. 12 - pp. 790, 685, 820,
                                           672, 783

A. Inhalation anthrax
B. Clostridial infection
C. Infectious abortion
D. Venereal disease
E. Diphtheria

711. ___ Bang's disease
712. ___ Chancroid
713. ___ Pseudomembrane
714. ___ Woolsorter's disease
715. ___ Anaerobic cellulitis   Ref. 12 - pp. 812, 799, 675,
                                              823, 836

ANSWER THE FOLLOWING QUESTIONS BY USING THE KEY
OUTLINED BELOW:
A. If 1 and 4 are correct
B. If 2 and 3 are correct
C. If 1, 2 and 3 are correct
D. If all are correct
E. If all are incorrect
F. If some combination other than the above is correct

716. BACTEROIDES ARE:
1. Normal inhabitants of the genital organs
2. Strictly anaerobic bacilli
3. Nonmotile
4. Gram-positive   Ref. 12 - p. 786

717. OF THE HEMOPHILUS, BOTH X AND V GROWTH FACTORS ARE
ESSENTIAL FOR:
1. H. parainfluenzae
2. H. influenzae
3. H. aegypticus
4. H. ducreyi   Ref. 12 - p. 790

718. BORDETELLA PERTUSSIS DIFFERS FROM HEMOPHILUS INFLUENZAE
AS FOLLOWS:
1. It produces only a single type of capsular antigen
2. It is rarely isolated from the blood stream
3. It does not require either X or V factor for growth after primary
   isolation
4. It requires only the X factor for growth
                              Ref. 12 - p. 795

SECTION III - BACTERIOLOGY

719. BARTONELLA BACILLIFORMIS:
1. Motile
2. Gram-negative
3. Coccobacillus
4. Peritrichous flagella. Ref. 12 - p. 918

720. THE NONPATHOGENIC SPECIES OF CORYNEBACTERIUM MOST OFTEN FOUND IN MAN:
1. C. hofmannii
2. C. ovis
3. C. kutscheri
4. C. xerosis Ref. 12 - p. 681

721. STAPHYLOCOCCAL FOOD POISONING:
1. Is characterized by sudden nausea
2. Occurs within a few hours after ingestion of contaminated food
3. Is a toxemia
4. Is an acute infection Ref. 12 - p. 737

722. MENINGOCOCCI AND GONOCOCCI:
1. Are difficult to cultivate
2. Ferment maltose
3. Can be differentiated by oxidase test
4. Tend to undergo rapid autolysis Ref. 12 - p. 743

723. GENETIC MATERIAL IS TRANSMITTED FROM ONE STRAIN OF ENTERIC BACILLUS TO ANOTHER BY:
1. Transduction
2. Lysogenization
3. Conjugation
4. Transformation Ref. 12 - p. 767

724. TYPHOID FEVER:
1. Acquired by ingestion of contaminated food
2. Incubation period is 7 to 14 days
3. Fever often increases in a step-like manner
4. "Rose spots" may appear on the trunk
  Ref. 12 - p. 775

725. SHIGELLAE:
1. Cause bacillary dysentery
2. Are motile
3. Produce gas during fermentation
4. Are less invasive than salmonellae Ref. 12 - p. 780

726. VIBRIO CHOLERAE:
1. Is pathogenic for man
2. Is shaped like a comma
3. Is gram-negative
4. Has peritrichous flagella Ref. 12 - p. 783

727. HEMOPHILUS INFLUENZAE:
1. Pleomorphic
2. Gram-positive
3. Iridescence increases with age
4. Virulent strains have capsules early in broth cultures
  Ref. 12 - p. 791

SECTION III - BACTERIOLOGY

728. THE FOLLOWING ANTIGENS OF B. ANTHRACIS HAVE BEEN AT LEAST PARTIALLY CHARACTERIZED:
1. The capsular polypeptide
2. The somatic polysaccharide
3. The protective antigen
4. The lipid surface antigen Ref. 12 - p. 822

729. THE FOLLOWING TYPES OF C. BOTULINUM TOXIN CAUSE HUMAN ILLNESS:
1. A
2. C
3. D
4. E Ref. 12 - p. 832

730. THE ENZYMES ELABORATED BY THE GROWING GAS GANGRENE ORGANISMS INCLUDE:
1. Collagenase
2. Proteinase
3. Deoxyribonuclease
4. Hyaluronidase Ref. 12 - p. 836

731. THE EXOTOXIN, θ-TOXIN, OF C. PERFRINGENS CROSS-REACTS IMMUNOLOGICALLY WITH:
1. Streptolysin O
2. Leukocidin
3. Staphylokinase
4. Tetanolysin Ref. 12 - p. 837

732. THE PURIFIED PROTEIN DERIVATIVE (PPD):
1. Is a more refined tuberculin
2. Is prepared by growing M. tuberculosis in a simple synthetic medium
3. Is precipitated with 50 percent saturated ammonium sulfate
4. Is a single protein without traces of nucleic acid and polysaccharides
  Ref. 12 - p. 855

733. THE FOLLOWING PHASES OF LEPROSY ARE DISTINGUISHED:
1. Lepromatous
2. Tuberculoid
3. Intermediate
4. Miliary Ref. 12 - p. 866

734. M. LEPRAE HAS A PREDILECTION FOR:
1. Skin
2. Blood
3. Lymphatics
4. Nerve Ref. 12 - p. 866

735. MYCOBACTERIA RESEMBLE MORE THE GRAM-NEGATIVE ORGANISMS THAN THE GRAM-POSITIVE ORGANISMS IN:
1. The high lipid content of the walls
2. The gram reaction
3. The basal glycopeptide layer in the walls
4. The protein content in their wall fraction
  Ref. 12 - p. 844

SECTION III - BACTERIOLOGY

AFTER EACH OF THE FOLLOWING CASE HISTORIES ARE SEVERAL
MULTIPLE CHOICE QUESTIONS BASED ON THE HISTORY. ANSWER
BY CHOOSING THE MOST APPROPRIATE ANSWER:

CASE HISTORY - QUESTIONS 736-740

A 20 year-old male soldier is suddenly taken ill with chills, vomiting, malaise and high fever. Several hours later he complains of a stiff neck and convulses. Petechiae are noted over the abdomen. That evening three more soldiers are admitted to the base hospital with the same symptoms.

736. THE MOST LIKELY DIAGNOSIS IS:
  A. Poliomyelitis
  B. Tuberculous meningitis
  C. Influenzal meningitis
  D. Meningococcal meningitis
  E. Tetanus
  Ref. 15 - p. 175

737. IF A SMEAR OF SPINAL FLUID FROM THE PATIENT REVEALED GRAM-NEGATIVE, SMALL, PLEOMORPHIC COCCOBACILLI, THEN THE PROBABLE ETIOLOGIC AGENT IS:
  A. Hemophilus influenzae
  B. Klebsiella pneumoniae
  C. Escherichia coli
  D. Proteus vulgaris
  E. Pasteurella pestis
  Ref. 15 - p. 212

738. ASSUMING THE AGENT ISOLATED FROM THE PATIENT'S SPINAL FLUID PRODUCES A POSITIVE OXIDASE TEST, THE MOST LIKELY DIAGNOSIS IS:
  A. Poliomyelitis
  B. Tuberculous meningitis
  C. Influenzal meningitis
  D. Meningococcal meningitis
  E. Tetanus
  Ref. 15 - p. 175

739. IF THE DIAGNOSIS RESTED BETWEEN INFLUENZAL MENINGITIS AND TUBERCULOUS MENINGITIS, WHICH OF THE FOLLOWING TESTS WOULD BE MOST USEFUL IN RAPID DIFFERENTIATION OF THE TWO?:
  A. PPD
  B. Cough plate
  C. Acid-fast stained smear of spinal fluid
  D. Guinea pig inoculation
  E. Bordet-Gengou rapid fixation method
  Ref. 15 - p. 191

740. OF THE DIAGNOSTIC CHOICES LISTED, WHICH ONE WOULD BENEFIT MOST FROM A PENICILLIN REGIMEN?:
  A. Poliomyelitis
  B. Tuberculous meningitis
  C. Influenzal meningitis
  D. Meningococcal meningitis
  E. Tetanus
  Ref. 15 - p. 175

CASE HISTORY - Questions 741-745

A 40 year-old male is brought to a hospital complaining of gradual onset of difficulty in swallowing. Several hours later the patient has 3 successive violent convulsions. The patient's wife states that her husband was wounded accidentally with a gunshot while duck hunting one week earlier.

741. THE PROBABLE ETIOLOGIC AGENT IS:
  A. A virulent virus
  B. Aerobic
  C. Invasive
  D. A toxin producer
  E. A gram-negative rod
  Ref. 6 - pp. 600-601

SECTION III - BACTERIOLOGY

742. ASSUMING THE ETIOLOGIC AGENT IS CLOSTRIDIUM TETANI, THE MICROSCOPIC APPEARANCE IS DESCRIBED AS:
    A. Drumstick
    B. Cuneiform
    C. Pleomorphic
    D. Serpentine
    E. Bipolar
    Ref. 6 - p. 600

743. MANY WOUNDED SOLDIERS DURING WORLD WAR I AND WORLD WAR II DEVELOPED A CONDITION KNOWN AS:
    A. Botulism
    B. Rabies
    C. Gas gangrene
    D. Bulbar poliomyelitis
    E. Encephalitis
    Ref. 6 - p. 590

744. FLORA OF GAS GANGRENE MAY INCLUDE:
    A. Cl. novyi
    B. Cl. septicum
    C. Cl. histolyticum
    D. Cl. perfringens
    E. All of the above
    Ref. 6 - p. 590

745. CLOSTRIDIUM PERFRINGENS IS ASSOCIATED WITH ALL OF THE FOLLOWING, EXCEPT:
    A. Stormy fermentation
    B. Anaerobiosis
    C. Non-motility
    D. Consistent spore formation
    E. Dirty wounds
    Ref. 6 - p. 593

CASE HISTORY - QUESTIONS 746-750

A 50 year-old businessman develops diarrhea 24 hours after leaving the Orient. The diarrhea is intense. The stools are thin and watery containing flakes. Stool culture reveals rapid growth of an almost pure culture at the surface of alkaline peptone broth. After 8 hours of incubation the colonies are small, colorless and translucent.

746. THE PROBABLE DIAGNOSIS IS:
    A. Typhoid fever
    B. Cholera
    C. Staphylococcus food poisoning
    D. Shigellosis
    E. Amebiasis
    Ref. 15 - p. 205

747. ASSUME A NON-MOTILE, GRAM-NEGATIVE ROD IS ISOLATED FROM THE STOOL. THE PROBABLE DIAGNOSIS WOULD BE:
    A. Typhoid fever
    B. Cholera
    C. Staphylococcus food poisoning
    D. Shigellosis
    E. Amebiasis
    Ref. 15 - p. 202

748. ASSUME THE DISEASE WAS TRANSMITTED BY INFECTED RAW EGGS. WHICH OF THE FOLLOWING ORGANISMS WOULD BE THE MOST LIKELY PATHOGEN?:
    A. Salmonella sp.
    B. Vibrio comma
    C. Staphylococcus aureus
    D. Shigella sonnei
    E. Entamoeba histolytica
    Ref. 15 - p. 201

749. ASSUMING THAT THE ETIOLOGIC AGENT DOES NOT HEMOLYZE GOAT RBC, IT WOULD BE:
    A. Salmonella typhosa
    B. Vibrio comma
    C. Staphylococcus aureus
    D. Shigella sonnei
    E. Entamoeba histolytica
    Ref. 6 - p. 551

750. ASSUMING THAT THE ORGANISM IS ONE OF THE "EL TOR" STRAINS:
    A. Hemolysis (sheep cells): negative
    B. Hemagglutination (chicken cells): negative
    C. Phage susceptibility (Mukerjee's phage IV): resistant
    D. Voges-Proskauer (Barritt method): negative
    E. Polymixin B: susceptible
    Ref. 6 - p. 551

SECTION III - BACTERIOLOGY   85

CASE HISTORY - Questions 751-755

A 3 year-old white female is brought to a doctor with the complaints of anorexia and sore throat. The child appears extremely ill. Physical examination reveals a gray membrane extensively involving the tonsils, uvula, soft palate and pharyngeal wall. Attempts to remove the membrane are followed by bleeding. The mother states that the girl has not received any immunization.

751. THE MOST LIKELY DIAGNOSIS:
   A. Infectious mononucleosis
   B. Herpetic tonsillitis
   C. Diphtheria
   D. Thrush
   E. Streptococcal tonsillitis
   Ref. 15 - p. 185

752. THE SCHICK TEST IS POSITIVE, THEREFORE:
   A. Circulating diphtheria antitoxin is present
   B. Immunity to diphtheria is absent
   C. Circulating scarlet fever antitoxin is present
   D. Circulating scarlet fever antitoxin is absent
   E. The infecting organism is non-toxigenic
   Ref. 15 - p. 186

753. ASSUME THE ETIOLOGIC ORGANISM IS NON-HEMOLYTIC AND OXIDIZES STARCH OR GLYCOGEN. THE ORGANISM IS:
   A. The gravis type of C. diphtheriae
   B. The mitis type of C. diphtheriae
   C. Streptococcus viridans
   D. Candida albicans
   E. Staphylococcus aureus
   Ref. 15 - p. 184

754. IF THE ORGANISM IS C. DIPHTHERIAE, THE FOLLOWING PROPERTIES ARE CHARACTERISTIC, EXCEPT:
   A. It is killed by boiling for 1 minute
   B. It is destroyed by the usual antiseptics
   C. It is moderately resistant to sulfonamides
   D. It is killed if kept at a temperature of $58°$ for 10 minutes
   E. It is resistant to penicillin
   Ref. 6 - p. 431

755. ASSUME THE DIAGNOSIS OF DIPHTHERIA IS CORRECT. THE PROPER EARLY TREATMENT SHOULD INCLUDE:
   A. Tracheotomy
   B. Acetic acid applications to the membrane
   C. Schick test
   D. Injections of antitoxin
   E. Vocal exercises
   Ref. 6 - p. 435

IDENTIFY THE ITEM IN EACH GROUP OF FIVE LISTED BELOW WHICH IS UNRELATED TO THE OTHERS:

756. SPIROCHETES:
   A. Multiply by transverse fission
   B. Have a well-defined nucleus
   C. Do not exhibit anterior-posterior polarity
   D. Are distinguishable from curved bacteria
   E. Are difficult to culture artificially   Ref. 2 - p. 745

757. PATHOGENIC SPIROCHETES:
   A. Borrelia
   B. Leptospira
   C. Cristispira
   D. Treponema pallidum
   E. Treponema pertunue
   Ref. 2 - p. 745

758. PRECIPITIN TESTS USED IN THE SERODIAGNOSIS OF SYPHILIS:
   A. Kahn test
   B. Kline test
   C. Hinton test
   D. Kolmer test
   E. Eagle test
   Ref. 2 - p. 755

759. CLINICAL CHARACTERISTICS OF RELAPSING FEVER:
   A. Gradual onset
   B. Chills
   C. Fever
   D. Headache
   E. Enlarged, tender spleen
   Ref. 2 - p. 747

760. RELAPSING FEVER MAY BE TRANSMITTED TO:
   A. Rat
   B. Mouse
   C. Guinea pig
   D. Human
   E. Monkey
   Ref. 2 - p. 747

761. LABORATORY DIAGNOSIS OF RELAPSING FEVER MAY BE MADE BY:
   A. Microscopic examination of blood smear
   B. Darkfield examination
   C. Animal inoculation and blood smear from animal examined microscopically
   D. Complement-fixation test
   E. Quantitative serology for reagin   Ref. 2 - p. 749

762. THE RELAPSING FEVER SPIROCHETE:
   A. Is a blood rather than a tissue parasite
   B. May be transmitted by blood sucking insects
   C. Is a tissue rather than a blood parasite
   D. May be transmitted by lice
   E. May be transmitted by ticks   Ref. 2 - p. 749

763. IN THE TICK VECTOR OF RELAPSING FEVER, THE SPIROCHETE:
   A. Is present in the coxal fluid
   B. Is present in the saliva
   C. Is present in the feces
   D. Is strain specific for a particular species of tick
   E. May be transmitted for several generations
                                       Ref. 2 - p. 749

764. VINCENT'S ANGINA ORGANISMS:
   A. Borrelia vincentii
   B. Cocci
   C. Bacilli
   D. Fusiform bacilli
   E. Treponema pallidum
   Ref. 2 - p. 750

765. PATHOGENIC TREPONEMA:
   A. Treponema pallidum
   B. Treponema caniculi
   C. Treponema pertenue
   D. Treponema carateum
   E. Treponema microdentium
   Ref. 2 - p. 750

766. SYPHILIS HAS BEEN TRANSMITTED TO:
   A. Rabbits
   B. Chimpanzee
   C. Dolphins
   D. Gibbons
   E. Monkeys
   Ref. 2 - p. 752

767. POSITIVE SERODIAGNOSTIC TESTS MAY BE OBTAINED IN:
   A. Venereal syphilis
   B. Non-venereal syphilis
   C. Yaws
   D. Pinta
   E. Quarta
   Ref. 2 - p. 752

SECTION III - BACTERIOLOGY

768. PERTAINING TO SYPHILIS:
    A. 10% of adults in the U.S.A. will give positive Wasserman tests
    B. 1-2% of children have congenital syphilis
    C. 2-29% of autopsies show evidence of syphilis
    D. It is the least intensively studied of the treponematoses
    E. In the U.S. syphilis is more prevalent in the black than in the white
    Ref. 2 - pp. 752-753

769. CHARACTERISTICS OF PRIMARY SYPHILIS:
    A. Indolent single ulcer
    B. Negative serology
    C. Positive darkfield examination
    D. The spirochete is limited to the primary site
    E. Initial lesion appears 10-30 days after contact
    Ref. 2 - pp. 753-754

770. SECONDARY SYPHILIS IS CHARACTERIZED BY:
    A. Constitutional symptoms
    B. Joint pain
    C. Onset 4 weeks or more after the chancre
    D. Eye invasion
    E. Inability to find spirochetes   Ref. 2 - p. 754

771. TERTIARY SYPHILIS IS CHARACTERIZED BY:
    A. Ulcerating skin lesions
    B. Gummata of internal organs
    C. Diverse signs of infection
    D. Persists for only a few months
    E. Spirochetes present only in small numbers
    Ref. 2 - p. 754

772. NEUROSYPHILIS IS CHARACTERIZED BY:
    A. General paresis
    B. Never occurs during secondary stage
    C. Spirochetal invasion of the central nervous system
    D. Tabes dorsalis
    E. Spirochetal invasion of pia mater   Ref. 2 - p. 754

773. CONGENITAL SYPHILIS:
    A. May result in abortion
    B. May go to term with the infant born dead
    C. In live births the lesions are those of tertiary syphilis
    D. Is characterized by many spirochetes in the tissues
    E. Is not generalized   Ref. 2 - p. 754

774. WASSERMAN ANTIBODY IS MEASURED IN:
    A. *T. pallidum* complement fixation test   D. Kline test
    B. Kahn test   E. Eagle test
    C. Hinton test   Ref. 2 - p. 755

775. FALSE NEGATIVE REACTIONS IN PATIENTS WITH SYPHILIS ARE DUE TO:
    A. Technical matters such as test sensitivity
    B. Early primary syphilis
    C. Late syphilis (latent)
    D. Lack of complement
    E. Presence of antibody to the Forssman antigen
    Ref. 2 - p. 756

SECTION III - BACTERIOLOGY

776. EFFECTIVE IN TREATMENT OF SYPHILIS:
    A. Terramycin
    B. Sulfonamides
    C. Penicillin
    D. Chloramphenicol
    E. Aureomycin
    Ref. 2 - p. 757

777. YAWS IS:
    A. Caused by <u>Treponema pertenue</u>
    B. Characterized by a papular eruption
    C. Characterized by the initial lesion being on the face
    D. Wasserman positive
    E. Venereal
    Ref. 2 - pp. 757-758

778. LEPTOSPIROSIS:
    A. May be manifested as Weil's disease
    B. May be acquired by contact with contaminated urine
    C. May be acquired by contact with infected tissue
    D. Organism is absent in urine
    E. Organism may be isolated from the blood
    Ref. 2 - p. 761

779. DISEASES CAUSED BY LEPTOSPIRA:
    A. Weil's disease
    B. Canicola fever
    C. Pomona fever
    D. Dengue fever
    E. Fort Bragg fever
    Ref. 2 - p. 762

780. SODUKU:
    A. Is caused by <u>Spirillum morsus muris</u>
    B. Has an incubation period of 10-22 days
    C. Is associated with fever and swollen lymph nodes
    D. Is caused by the scratch of a young cat
    E. Has a relapsing course
    Ref. 2 - p. 765

FOR EACH OF THE FOLLOWING MULTIPLE CHOICE QUESTIONS CHOOSE THE <u>ONE</u> MOST APPROPRIATE ANSWER:

781. T. PALLIDUM IS <u>NOT</u> CHARACTERIZED BY:
    A. A flexible body 5-20 micra in length
    B. Pointed ends
    C. Regular sharp spirals
    D. A coiled contractile axial filament
    E. Slow rotation
    Ref. 6 - p. 643

782. THE WASSERMAN REACTION IS BASED ON MEASUREMENT OF:
    A. T. P. I. antibodies
    B. Reagin
    C. Agglutinating antibodies
    D. Complement-fixing antibodies
    E. Gamma globulin
    Ref. 6 - p. 644

783. T. PALLIDUM PRODUCES:
    A. Exotoxin
    B. Endotoxin
    C. Enterotoxin
    D. Erythrogenic toxin
    E. None of the above
    Ref. 6 - p. 645

784. MOST COMMON SITE OF PRIMARY LESION OF SYPHILIS:
    A. Lips
    B. Soft palate
    C. Genitalia
    D. Breasts
    E. Inguinal lymph nodes
    Ref. 6 - p. 645

SECTION III - BACTERIOLOGY

785. FOLLOWING TREATMENT WITH PENICILLIN, T. PALLIDUM USUALLY DISAPPEARS FROM THE LOCAL PRIMARY LESION WITHIN:
    A. 6-24 hours
    B. 24-48 hours
    C. One week
    D. 2-4 weeks
    E. Six months
    Ref. 6 - p. 646

786. THE PRIMARY LESION OF SYPHILIS WILL:
    A. Become painful and indurated
    B. Heal spontaeously, even if untreated
    C. Become a chronic chancre
    D. Cause gangrene of the penis
    E. None of the above
    Ref. 6 - p. 646

787. DURING SECONDARY PHASE OF SYPHILIS, THERE IS:
    A. Recurrent chancre formation
    B. Gumma formation
    C. Multiplicity of lesions
    D. Tabes dorsalis
    E. General paresis
    Ref. 6 - p. 646

788. SEROLOGIC TESTS FOR SYPHILIS ARE PRACTICALLY ALWAYS POSITIVE DURING:
    A. The incubation period
    B. Primary syphilis
    C. Secondary syphilis
    D. Late latent syphilis
    E. Tertiary syphilis
    Ref. 6 - p. 646

789. TERTIARY SYPHILIS IS NOT CHARACTERIZED BY:
    A. Multiple lesions of skin and mucous membrane
    B. Gumma formation
    C. General paresis
    D. Tabes dorsalis
    E. Cardiovascular disease
    Ref. 6 - p. 646

790. THE HIGHEST INCIDENCE OF FALSE POSITIVE SEROLOGIC TESTS FOR SYPHILIS OCCURS IN:
    A. Rheumatoid arthritis
    B. Leptospirosis
    C. Disseminated lupus erythematosus
    D. Malaria
    E. Advanced tuberculosis
    Ref. 6 - p. 647

791. NEARLY ALL SYPHILITIC INFECTIONS ARE ACQUIRED BY:
    A. Sexual contact
    B. Kissing
    C. Abnormal sexual practices
    D. All of the above
    E. None of the above
    Ref. 6 - p. 648

792. RELAPSING FEVER IS CAUSED BY:
    A. Borrelia buccale
    B. Borrelia relapsis
    C. Borrelia recurrentis
    D. Treponema pertenue
    E. None of the above
    Ref. 6 - p. 652

793. THE CHIEF VECTOR OF RELAPSING FEVER IS:
    A. Mite
    B. Soft tick
    C. Hard tick
    D. Mosquito
    E. Fly
    Ref. 6 - p. 654

794. THE ORGANISM NOT FOUND IN ASSOCIATION WITH THE LESION OF VINCENT'S ANGINA:
    A. Fusiform bacilli
    B. Spirochetes
    C. Cocci
    D. Vibrios
    E. Lactobacilli
    Ref. 6 - p. 655

SECTION III - BACTERIOLOGY

795. SOME OF THE ORGANISMS CAUSING VINCENT'S ANGINA PROBABLY PRODUCE:
    A. Exotoxins
    B. Endotoxins
    C. Proteolytic enzymes
    D. Hemolysins
    E. None of the above
    Ref. 6 - p. 656

796. LEPTOSPIRA ORGANISMS ARE NOT CHARACTERIZED BY:
    A. Motility
    B. Aerobic growth
    C. An external axistyle
    D. Optimal growth at pH 9.0
    E. Absence of flagella
    Ref. 6 - p. 660

797. SPIROCHETES ARE SUSCEPTIBLE TO:
    A. Arsenic and antimony
    B. Bismuth
    C. Mercury compounds
    D. Penicillin
    E. All of the above
    Ref. 6 - p. 638

798. THE DESTRUCTIVE NATURE OF GUMMAS IS ATTRIBUTED TO:
    A. Tolerance
    B. Hypersensitivity
    C. Reagins
    D. Antitoxins
    E. Agglutinins
    Ref. 6 - p. 646

799. SYSTEMIC USE OF CORTISONE THERAPY FOR SYPHILIS MAY CAUSE:
    A. The organisms to become penicillin resistant
    B. Increased T.P.I. antibody formation and rapid cure
    C. Decreased T.P.I. antibody and increased spirochetemia
    D. False positive serologic tests
    E. Increased number of sexual contacts because of increased libido
    Ref. 6 - p. 648

800. THE COMMON FEATURE OF BEJEL, PINTA AND YAWS:
    A. They are attenuated forms of syphilis
    B. They are all venereal diseases
    C. They are resistant to penicillin therapy, unlike syphilis
    D. They are all caused by spirochetes
    E. They are all acute diseases of the New World
    Ref. 6 - p. 648

ANSWER THE FOLLOWING QUESTIONS BY USING THE KEY OUTLINED BELOW:
A. If both statement and reason are true and related cause and effect
B. If both statement and reason are true but not related cause and effect
C. If the statement is true but the reason is false
D. If the statement is false but the reason is true
E. If both statement and reason are false

801. Spirochetes are found widely in nature BECAUSE they are motile organisms. Ref. 6 - p. 638

802. The spirochetes in the local primary lesion are believed to be destroyed by local tissue immunity rather than by humoral immunity BECAUSE an unhindered multiplication of the organisms in the skin and the mucous membranes preparatory to the explosive onset of the secondary lesions occurs simultaneously with the elimination of the spirochetes from the local lesion. Ref. 6 - p. 646

803. There is great danger of transmitting syphilis by blood transfusion from refrigerated banked blood BECAUSE blood is a good culture medium for this organism. Ref. 6 - p. 644

SECTION III - BACTERIOLOGY

804. The patient with syphilis cannot transmit the disease before the development of a primary lesion BECAUSE the incubation period of syphilis averages about 3 weeks. Ref. 6 - pp. 645-646

805. If the mother has been adequately treated for syphilis during pregnancy, the child will be born free of syphilis BECAUSE the mother transmits sufficient level of protective antibodies to the child.
Ref. 6 - p. 647

806. Spirochetes are rarely demonstrated in the tissues of children with congenital syphilis BECAUSE the organisms are unable to pass the placental barrier. Ref. 6 - p. 647

807. Jarisch-Herxheimer reactions occur following the beginnings of penicillin treatment BECAUSE this phenomenon is believed to be an allergic reaction which is precipitated by the sudden release of antigen from the spirochetes that are killed by penicillin. Ref. 6 - p. 648

808. Protection of the skin from invasion by Leptospira icterohaemorrhagiae is not important BECAUSE the organism cannot enter the body through an abrasion on the skin. Ref. 6 - pp. 663, 664

809. The diseases produced by Spirillum minus and Streptobacillus moniliformis are both called rat-bite fever BECAUSE both can be transmitted by the bite of a rat. Ref. 6 - p. 666

810. The diagnosis of rat-bite fever caused by Spirillum minus can be confirmed by inoculating white mice or guinea pigs BECAUSE rat-bite fever begins as a wound infection. Ref. 6 - pp. 666, 667

811. Mycoplasmas are widespread in nature BECAUSE they are intermediate in size between bacteria and viruses. Ref. 6 - p. 630

812. The growth requirements of mycoplasmas are fastidious BECAUSE they possess a cell wall. Ref. 6 - p. 630

813. Mycoplasmas are part of the normal flora of the body BECAUSE they have been isolated so frequently from the oral cavity and from the urinary tract in the absence of inflammation. Ref. 6 - p. 630

814. Little is known about mutations of T. pallidum BECAUSE only in vivo methods are available for studying variations of the organism.
Ref. 12 - p. 886

815. Second attacks of mycoplasmal pneumonia have not been reported BECAUSE at least three kinds of antibody are formed as a result of mycoplasmal infection. Ref. 12 - pp. 909-910

SECTION III - BACTERIOLOGY

ANSWER THE FOLLOWING QUESTIONS BY USING THE KEY OUTLINED BELOW:
A. If 1 and 4 are correct
B. If 2 and 3 are correct
C. If 1, 2 and 3 are correct
D. If all are correct
E. If all are incorrect
F. If some combination other than the above is correct

816. MYCOPLASMAS ARE RESISTANT TO THE ANTIMICROBIAL ACTIONS OF:
1. Sulfonamides
2. Tetracyclines
3. Kanamycin
4. Penicillin
Ref. 12 - p. 902

817. MYCOPLASMA PNEUMONIAE DIFFERS FROM THE OTHER SPECIES OF MYCOPLASMA IN ITS:
1. Slowness of growth
2. Production of a rapid hemolysin for guinea pig red cells
3. Ability to hemadsorb erythrocytes of many animal species
4. Reduction of tetrazolium under aerobic conditions
Ref. 12 - p. 904

818. THE FOLLOWING SPECIES OF MYCOPLASMA ARE FOUND IN THE GENITOURINARY TRACT:
1. M. salivarium
2. M. fermentans
3. M. hominis
4. M. orale
Ref. 12 - pp. 903-904

819. THE BEDSONIAE ARE NO LONGER CONSIDERED AS VIRUSES BECAUSE THEY:
1. Possess both DNA and RNA
2. Possess wall components characteristic of bacteria
3. Possess a variety of biosynthetic enzymes
4. Are larger than any of the viruses
Ref. 12 - p. 1014

820. THE FOLLOWING PROPERTIES OF RICKETTSIAE INDICATE THAT THEY ARE BACTERIA:
1. They multiply by binary fission
2. They contain both DNA and RNA
3. They possess a number of indigenous enzymes
4. Their growth is inhibited by a variety of antibacterial agents
Ref. 12 - p. 932

821. THE DIFFERENCES AMONG THE VARIOUS RICKETTSIAL DISEASES ARE FOUND IN:
1. The arthropod vector
2. The intracellular localization of the organism
3. Distribution and appearance of rash
4. The specificities of the antibodies formed
Ref. 12 - p. 934

822. ROCKY MOUNTAIN SPOTTED FEVER:
1. The organism grows within the nuclei as well as in the cytoplasm
2. Onset of disease occurs 1 to 2 weeks after tick bite
3. Rash begins peripherally on the ankles, wrists, and forehead and then spreads to the trunk
4. Does not respond to chloramphenicol and tetracyclines
Ref. 12 - p. 940

SECTION III - BACTERIOLOGY

823. R. AKARI DIFFERS FROM THE OTHER SPOTTED FEVER AGENTS IN THAT:
    1. It is transmitted by a mite
    2. It is transmitted by a tick
    3. It elicits antibodies to Proteus-OX strains
    4. It does not elicit antibodies to Proteus-OX strains
    Ref. 12 - p. 941

824. COXIELLA BURNETII DIFFERS FROM OTHER RICKETTSIAE IN THE FOLLOWING RESPECTS:
    1. It is unstable outside host cells
    2. Infection is acquired by inhalation of contaminated particles
    3. Disease in man is usually characterized by pneumonitis, without a rash
    4. It elicits antibodies for only the OX strains of Proteus
    Ref. 12 - p. 943

825. TRENCH FEVER:
    1. Is characterized by an abrupt onset
    2. Is often called 5-day fever
    3. Rash is commonly present during the febrile periods
    4. Is rarely severe
    Ref. 12 - p. 944

AFTER EACH OF THE FOLLOWING CASE HISTORIES ARE SEVERAL MULTIPLE CHOICE QUESTIONS BASED ON THE HISTORY. ANSWER BY CHOOSING THE MOST APPROPRIATE ANSWER:

CASE HISTORY - Questions 826-830

A 3 year-old boy is bitten on the arm by a rat. Seven days later the bite mark ulcerates and a rash and fever become apparent. A smear of the lesion reveals spiral bacteria.

826. THE MOST PROBABLE ORGANISM IS:
    A. Vibrio fetus
    B. Streptobacillus moniliformis
    C. Spirillum volutans
    D. Spirillum minus
    E. Leptospira icterohaemorrhagiae   Ref. 6 - p. 667

827. ASSUMING THE ETIOLOGIC ORGANISM IS STREPTOBACILLUS, THE DIAGNOSIS IS:
    A. Relapsing fever   D. Canicola fever
    B. Rat-bite fever   E. Swineherd's fever
    C. Weil's disease   Ref. 6 - p. 614

828. ASSUME THE CHILD DEVELOPED JAUNDICE AND SKIN HEMORRHAGES AFTER SWIMMING IN RAT-CONTAMINATED WATER. THE MOST PROBABLE DIAGNOSIS WOULD THEN BE:
    A. Relapsing fever   D. Swineherd's fever
    B. Rat-bite fever   E. Weil's disease
    C. Canicola fever   Ref. 6 - p. 663

829. IF THE ORGANISM IS TIGHTLY COILED, THIN, AND FLEXIBLE, IT IS:
    A. Spirillum minus
    B. Vibrio fetus
    C. Borrelia recurrentis
    D. Treponema pertenue
    E. Leptospira icterohaemorrhagiae   Ref. 6 - p. 660

SECTION III - BACTERIOLOGY

830. IF THE DIAGNOSIS IS INDEED RAT-BITE FEVER BY <u>SPIRILLUM MINUS</u>, **THE** DRUG OF CHOICE IS:
   A. Penicillin
   B. Sulfonamides
   C. Streptomycin
   D. All of the above
   E. None of the above
   Ref. 6 - p. 667

CASE HISTORY - Questions 831-835

A 10 year-old boy is admitted to a teaching hospital because of severe chills, headache, joint pains and myalgia of 24 hours duration. WBC is 13,000. The temperature is 104° F.

831. THE DIFFERENTIAL DIAGNOSIS SHOULD INCLUDE:
   A. Influenza
   B. Typhus fever
   C. Rocky Mountain spotted fever
   D. Pneumococcal pneumonia
   E. All of the above
   Ref. 6 - pp. 368, 680, 686-687, 898

832. FURTHER HISTORY REVEALS THAT THE CHILD WAS ON A CAMPING TRIP IN THE ROCKY MOUNTAIN AREA ONE WEEK PRIOR TO THE ONSET OF SYMPTOMS. ROCKY MOUNTAIN SPOTTED FEVER WOULD BE A LIKELY DIAGNOSIS IF A RASH APPEARED:
   A. At the time of onset of symptoms
   B. 2-5 days after onset
   C. 1-2 weeks after onset
   D. More than 2 weeks after onset
   E. Not at all
   Ref. 6 - p. 687

833. ASSUME THAT THIS PATIENT HAS ROCKY MOUNTAIN SPOTTED FEVER. IT HAS BEEN TRANSMITTED BY:
   A. Mite bite
   B. Flea feces
   C. Louse feces
   D. Tick bite
   E. Flea bite
   Ref. 6 - p. 687

834. THREE WEEKS AFTER THE ONSET OF ROCKY MOUNTAIN SPOTTED FEVER, THE FEVER IS DOWN AND THERE IS A FOUR-FOLD RISE IN AGGLUTININS TO <u>PROTEUS OX19</u>. THE RISE IN TITER INDICATES:
   A. The infection has not reached its peak
   B. The prognosis is very poor
   C. The patient is in convalescence
   D. The patient is completely recovered
   E. He has developed complete immunity
   Ref. 6 - p. 687

835. THE TREATMENT OF CHOICE FOR ROCKY MOUNTAIN SPOTTED FEVER IS:
   A. Hyperimmune rabbit serum
   B. Para-aminobenzoic acid
   C. Penicillin
   D. Chlortetracycline
   E. Sulfonamides
   Ref. 6 - p. 688

CASE HISTORY - Questions 836-840

A 45 year-old woman is admitted to the hospital following the abrupt onset of chills, fever, headache and prostration. She has a non-productive cough. WBC is normal. Her husband states that her pet parakeet died three days before the onset of the patient's illness.

836. THE DIAGNOSIS THAT IS MOST STRONGLY SUGGESTED:
   A. Tuberculosis
   B. Psittacosis
   C. Rickettsialpox
   D. Q fever
   E. Epidemic typhus
   Ref. 6 - p. 708

SECTION III - BACTERIOLOGY   95

837. IF THE PATIENT HAS RICKETTSIALPOX, INITIAL PHYSICAL EXAMINATION SHOULD REVEAL:
A. A diffuse papular rash
B. A tick attached to the skin
C. A red papule with enlarged tender regional lymph nodes
D. Hepato-splenomegaly
E. No specific abnormalities   Ref. 6 - p. 689

838. IF THE DISEASE IS RICKETTSIALPOX, IT HAS BEEN TRANSMITTED BY:
A. Mite bite   D. Tick bite
B. Flea feces   E. Flea bite
C. Louse feces   Ref. 6 - p. 689

839. IF THE PATIENT HAS PSITTACOSIS:
A. The respiratory tract is the main portal of entry
B. He must have been exposed to an infected parrot
C. The mode of transmission was by a tick bite
D. Hematologic findings are specific
E. Agglutinins to Proteus OX19 are present
               Ref. 6 - p. 708

840. IF THE PATIENT HAS Q FEVER:
A. Agglutinins to Proteus OX19 or OX2 can be demonstrated
B. Cross-immunity to other rickettsial diseases will develop
C. Complement-fixing antibodies in a titer of more than 1:20 will be demonstrated
D. A skin rash will develop during the first week
E. A fatal outcome is quite likely   Ref. 6 - p. 699

CASE HISTORY - Questions 841-845

The 6 year-old son of an American Indian is seen in the office with swelling and ptosis of the upper lid of one eye. There is marked conjunctival inflammation, papillary hypertrophy, and superficial reddening of the cornea.

841. THE MOST SERIOUS DISEASE TO BE CONSIDERED IS:
A. Inclusion conjunctivitis   D. Epidemic keratoconjunctivitis
B. Trachoma   E. Psittacosis
C. Ophthalmia neonatorum   Ref. 6 - pp. 703-704

842. INCLUSION CONJUNCTIVITIS CAN BE TENTATIVELY DIFFERENTIATED FROM TRACHOMA BY:
A. Demonstration of inclusion bodies
B. More marked involvement of the upper lid
C. More marked involvement of the lower lid
D. Equal involvement of the lower lid
E. The early onset of blindness   Ref. 15 - p. 277

843. IN TRACHOMA, THE INCLUSION BODIES FROM EPITHELIAL CELLS MAY BE:
A. Stained with Giemsa
B. Stained with fluorescent antibody
C. Inoculated into yolk sac
D. Inoculated into tissue culture
E. All of the above   Ref. 6 - p. 704

SECTION III - BACTERIOLOGY

844. TOPICAL CORTISONE THERAPY IS NOT USED IN EITHER TRACHOMA OR INCLUSION CONJUNCTIVITIS BECAUSE:
   A. A prompt cure may result
   B. Broad spectrum antibiotics are more effective
   C. Sulfonamide therapy is more effective
   D. It is too expensive to use in such a widespread disease
   E. Reactivation of a subsiding infection may occur
   Ref. 15 - p. 277

845. THE COMMON MODE OF TRANSMISSION OF TRACHOMA AND INCLUSION CONJUNCTIVITIS:
   A. By a newborn infant during birth
   B. Directly from eye to eye by fingers
   C. By sexual contact
   D. By swimming in unchlorinated swimming pools
   E. All of the above
   Ref. 15 - p. 277

CASE HISTORY - Questions 846-850

You are a military doctor on duty with the U.S. Army in the South Pacific during the Second World War. An asymptomatic soldier is seen on sick call with an eschar on his right leg and inguinal adenopathy on the same side. He states that the eschar is at the site of a recent mite bite. Your tentative diagnosis is tsutsugamushi disease.

846. YOU SHOULD:
   A. Treat the eschar with iodine, give him a shot of penicillin, and send him back to duty
   B. Call his commanding officer and tell him the man is a goldbrick
   C. Call his commanding officer and tell him the man will have to be hospitalized for at least one month
   D. Call his commanding officer and tell him to send the man to the front lines because the mortality rate with this disease approaches 100% and he might as well die in the defense of his country
   E. Draw blood for agglutinins to Proteus OX19 and OXK
   Ref. 6 - pp. 694-695

847. ASSUME YOUR DIAGNOSIS IS CORRECT. THE ONLY OTHER RICKETTSIAL DISEASE TRANSMITTED BY THE MITE IS:
   A. Epidemic typhus
   B. Rocky Mountain spotted fever
   C. Trench fever
   D. Rickettsialpox
   E. Q fever
   Ref. 6 - p. 689

848. THIS MAN'S DISEASE CAN BE DIFFERENTIATED FROM THE OTHER RICKETTSIAL DISEASES BECAUSE:
   A. No other rickettsial disease is found in the South Pacific
   B. This man will develop rising titer of agglutinins to Proteus OX19
   C. This man will develop rising titer of agglutinins to Proteus OXK
   D. This is the only rickettsial disease that responds to chloramphenicol
   E. This is the only rickettsial disease that does not cause a rash
   Ref. 6 - p. 695

SECTION III - BACTERIOLOGY   97

849. WHICH STATEMENT IS CORRECT FOR TSUTSUGAMUSHI RICKETTSIOSES?:
   A. The disease was named after the Japanese who discovered the etiologic agent
   B. The rickettsia grows very poorly in the yolk sac of the embryonated chicken egg
   C. There is cross-immunity between this disease and other rickettsial diseases
   D. The disease can be prevented by spraying larval nesting places with DDT
   E. The rickettsia, R. tsutsugamushi, grows in the cytoplasm of the infected cells
   Ref. 6 - p. 694

850. IF YOUR DIAGNOSIS WAS INCORRECT AND THE MAN HAS MURINE (ENDEMIC) TYPHUS, YOU CAN EXPECT:
   A. The man to develop agglutinins to Proteus OX19 and OX2
   B. The man will not develop agglutinins to Proteus
   C. The mortality rate to be higher than you anticipated
   D. A law suit for malpractice to be instituted by his survivors
   E. None of the above   Ref. 6 - p. 683

CASE HISTORY - Questions 851-855

You are called as a consultant to see a patient with chills, weakness, severe headache, and temperature of 104°. Five days after the onset of symptoms, he developed a rash composed of pink spots which later became hemorrhagic. The WBC is normal. The presumptive diagnosis is epidemic typhus.

851. THE DIAGNOSIS MAY BE CONFIRMED BY:
   A. Determination of agglutinins to Proteus OX19 and OXK
   B. Determination of the endotoxin produced by the organism
   C. The clinical picture and normal WBC
   D. Demonstration of the rickettsiae in yolk sac smears following inoculation of infected guinea pig brain into yolk sac of embryonated egg
   E. The absence of agglutinins to Proteus OX19 and OXK
   Ref. 6 - pp. 680, 682

852. TYPICALLY, THE RASH IN EPIDEMIC TYPHUS:
   A. Is most severe on the face
   B. Starts on the trunk and spreads to the extremities
   C. Starts on the extremities and spreads to the trunk
   D. Rarely occurs on the extremities
   E. Occurs simultaneously with the onset of illness
   Ref. 6 - p. 680

853. THIS DISEASE IS TRANSMITTED BY:
   A. Ticks   D. Rats
   B. Fleas   E. Mites
   C. Lice   Ref. 6 - p. 680

854. THIS DISEASE OCCURS MOST FREQUENTLY IN EPIDEMIC FORM FOLLOWING:
   A. Long sea voyages
   B. Vacations in endemic areas
   C. The winter months
   D. The unsanitary disposal of human feces
   E. Wars   Ref. 6 - p. 678

SECTION III - BACTERIOLOGY

855. ALTHOUGH THE RICKETTSIA MAY DEVELOP RESISTANCE TO THE DRUG, THE MOST EFFECTIVE TREATMENT FOR EPIDEMIC TYPHUS IS:
   A. Penicillin
   B. Chloramphenicol
   C. Para-aminobenzoic acid
   D. Hyperimmune serum
   E. Sulfonamides
   Ref. 6 - p. 681

FOR EACH OF THE FOLLOWING MULTIPLE CHOICE QUESTIONS. CHOOSE THE ONE MOST APPROPRIATE ANSWER:

856. THE CHLAMYDIAE DIFFER FROM TRUE VIRUSES IN THE FOLLOWING CHARACTERISTICS, EXCEPT:
   A. They possess both RNA and DNA
   B. They multiply by binary fission
   C. They possess bacterial type cell walls
   D. They possess ribosomes
   E. They possess only RNA
   Ref. 15 - p. 271

857. WHICH ONE OF THE FOLLOWING STATEMENTS DOES NOT DESCRIBE TRACHOMA?:
   A. Under natural conditions, it is found only in man
   B. Only epithelial cells of the conjunctiva are infected
   C. The infectivity of the organism for chick embryo cells is extremely high
   D. It is spread by contact
   E. It is prevalent in Egypt, North Africa and the Near East
   Ref. 12 - p. 957

858. CULTIVATION OF THE FOLLOWING RICKETTSIA ON A CELL-FREE MEDIUM HAS RECENTLY BEEN REPORTED FOR THE FIRST TIME:
   A. R. prowazekii
   B. R. mooseri
   C. R. rickettsii
   D. C. burnetii
   E. R. quintana
   Ref. 12 - p. 928

859. WALL PREPARATIONS FROM THE FOLLOWING SPECIES HAVE BEEN SHOWN TO CONTAIN MURAMIC ACID AS IN BACTERIA:
   A. R. prowazekii
   B. R. mooseri
   C. R. rickettsii
   D. R. conori
   E. C. burnetii
   Ref. 12 - p. 929

860. RICKETTSIALPOX:
   A. The causative agent multiplies only in the nucleus
   B. The causative agent multiplies only in the cytoplasm
   C. Its particulate type-specific antigen does not cross-react with R. rickettsii in complement-fixation assays
   D. The house mouse constitutes the natural mammalian reservoir for the causative agent
   E. Lymph nodes are not enlarged
   Ref. 12 - p. 941

861. THE FOLLOWING STATEMENTS DESCRIBE THE RICKETTSIAE, EXCEPT:
   A. They are small
   B. They are obligately parasitic
   C. They possess a typical bacterial cell wall
   D. They all produce fever and rash in man
   E. They possess most of the enzymes of bacteria
   Ref. 15 - p. 266

862. COXIELLA BURNETII IS THE CAUSATIVE AGENT OF:
   A. Tsutsugamushi disease
   B. Rocky Mountain spotted fever
   C. Murine typhus
   D. Q fever
   E. Rickettsialpox
   Ref. 2 - p. 833

SECTION III - BACTERIOLOGY

863. FLEAS ARE THE ARTHROPOD VECTORS OF:
   A. Brill's disease
   B. Kala-azar
   C. Q fever
   D. Epidemic typhus
   E. Endemic typhus
   Ref. 2 - p. 832

864. THE CAUSATIVE AGENT OF Q FEVER IS:
   A. Filterable
   B. Easily stained with basic aniline dyes
   C. Unknown
   D. R. typhi
   E. 5 µ in length
   Ref. 2 - p. 841

865. THE SERUM FROM PATIENTS WITH THE FOLLOWING DISEASE DOES NOT GIVE A POSITIVE WEIL-FELIX REACTION:
   A. Epidemic typhus
   B. Endemic typhus
   C. Q fever
   D. Scrub typhus
   E. Rocky Mountain spotted fever
   Ref. 15 - p. 268

866. ORDINARY MILK PASTEURIZATION MAY NOT ELIMINATE:
   A. Corynebacterium diphtheriae
   B. Streptococcus pyogenes
   C. Coxiella burnetii
   D. Mycobacterium tuberculosis
   E. None of the above
   Ref. 2 - p. 842

867. RICKETTSIAE GROW BEST IN:
   A. Blood-agar
   B. Bordet-Gengou medium
   C. Tissue cultures
   D. Chick embryo yolk sac linings
   E. Liquid media
   Ref. 2 - p. 827

868. PEDICULUS HUMANUS IS THE ARTHROPOD VECTOR OF:
   A. Bubonic plague
   B. Tsutsugamushi disease
   C. African sleeping sickness
   D. Epidemic typhus
   E. Rocky Mountain spotted fever
   Ref. 2 - p. 832

869. THE ARTHROPOD VECTOR OF TSUTSUGAMUSHI DISEASE IS A:
   A. Flea
   B. Tick
   C. Mite
   D. Louse
   E. Fly
   Ref. 2 - p. 833

870. THE MORPHOLOGY OF RICKETTSIAE:
   A. Bacillus
   B. Coccus
   C. Diplococcus
   D. Filamentous
   E. Varied
   Ref. 2 - p. 826

871. R. TYPHI IS THE CAUSATIVE AGENT OF:
   A. Relapsing fever
   B. Q fever
   C. Epidemic typhus
   D. Endemic typhus
   E. Tularemia
   Ref. 2 - p. 832

872. ALL RICKETTSIAE ARE:
   A. Human pathogens exclusively
   B. Poorly-established parasites of arthropods
   C. Intracellular parasites
   D. Spore formers under unfavorable conditions
   E. All of the above
   Ref. 2 - p. 826

SECTION III - BACTERIOLOGY

873. BRILL'S DISEASE REFERS TO:
 A. Rickettsial infection resembling a follicular lymphoma
 B. Recrudescent typhus
 C. Q fever
 D. Cat-scratch fever
 E. Rickettsialpox   Ref. 2 - p. 835

874. TICKS ARE THE ARTHROPOD VECTORS OF:
 A. Tsutsugamushi disease   D. Endemic typhus
 B. Bubonic plague   E. Epidemic typhus
 C. Rocky Mountain spotted fever Ref. 2 - p. 832

875. R. RICKETTSII IS THE CAUSATIVE AGENT OF:
 A. Tsutsugamushi disease   D. Q fever
 B. Rocky Mountain spotted fever E. Rickettsialpox
 C. Murine typhus   Ref. 2 - p. 832

SELECT THE ITEM FROM COLUMN I WHICH IS UNRELATED TO THE FOUR OTHER ITEMS IN THIS COLUMN. CIRCLE THIS NUMBER. IDENTIFY THE ITEM IN COLUMN II WITH WHICH THE FOUR RELATED ITEMS IN COLUMN I ARE ASSOCIATED. CIRCLE THIS LETTER:

                I   II

876. 1. Rickettsia prowazeki   A. Gram variable rickettsiae
 2. Coxiella burnetii   B. Gram-negative rickettsiae
 3. Rickettsia tsutsugamushi   C. Gram-positive rickettsiae
 4. Rickettsia ricketsii
 5. Bartonella bacilliformis   Ref. 2 - pp. 825, 832-833

877. 1. Rickettsiae   A. Intracellular growth only
 2. ECHO viruses   B. Extracellular growth only
 3. Coxsackie viruses   C. Intracellular and extra-
 4. Adenoviruses   cellular growth
 5. Gonococcus   Ref. 2 - pp. 825-826

878. 1. Dermacentor andersoni   A. Rickettsia is pathogenic
 2. Louse   for these vectors
 3. Mite   B. Rickettsia is non-pathogenic
 4. Dermacentor variabilis   for these vectors
 5. Flea   C. Not a vector for rickettsia
                                          Ref. 2 - p. 828

879. 1. Increases vascular permeability   A. Endotoxin only of rickettsiae
 2. Intimately associated with the   B. Exotoxin only of rickettsiae
    rickettsiae   C. Endotoxin and exotoxin of
 3. Lethal for mice   rickettsiae
 4. Unaffected by ultraviolet light
 5. Not neutralized by antiserum   Ref. 2 - pp. 828-829

880. 1. Rocky Mountain spotted fever   A. Typhus group
 2. Mediterranean fever   B. Spotted fever group
 3. Queensland tick typhus   C. Scrub typhus
 4. Endemic typhus
 5. Rickettsialpox   Ref. 15 - p. 266

881.
1. Rickettsia rickettsii
2. Rickettsia typhi
3. Rickettsia australis
4. Rickettsia conori
5. Rickettsia akari

A. Cause typhus-like disease
B. Cause spotted fever-like disease
C. Cause tsutsugamushi-like disease

Ref. 2 - p. 834

882.
1. Initial violent headache
2. Macular eruption
3. 1-2 day incubation
4. Gangrene
5. Bronchopneumonia

A. Salmonellosis
B. Typhus
C. Meningococcus

Ref. 2 - p. 834

883.
1. Typhus
2. Rocky Mountain spotted fever
3. Q fever
4. Brill's disease
5. Psittacosis

A. Latent infections
B. New acute infections
C. Rickettsial diseases of man

Ref. 2 - pp. 832-833

884.
1. Rickettsia prowazeki
2. Rickettsia typhi
3. Rickettsia rickettsii
4. Rickettsia tsutsugamushi
5. Rickettsia akari

A. Vertebrate reservoir other than man
B. Positive Weil-Felix for OX19, OX2
C. Single species of insect vector

Ref. 2 - pp. 832-833

885.
1. Weigl vaccine (infected lice)
2. Vaccine prepared from yolk sac culture
3. Destruction of lice with DDT
4. Mouse brain vaccine
5. Rat lung vaccine

A. Theoretically effective in the prevention of typhus
B. Used in immunization against typhus
C. Used in active immunization against Q fever

Ref. 2 - p. 837

886.
1. Idaho, Montana, Wyoming, Oregon and Washington
2. South Atlantic states
3. Ticks
4. Lice
5. Case fatality rate of 18-19%

A. Rickettsia typhi
B. Rickettsia akari
C. Rickettsia rickettsii

Ref. 2 - p. 838

887.
1. Boutonneuse fever
2. North Queensland tick typhus
3. Siberian tick typhus
4. South African tick fever
5. Q fever

A. Closely related to the spotted fevers
B. Closely related to typhus
C. Closely related to scrub typhus

Ref. 2 - p. 839

888.
1. First discovered in New York City
2. Varicelliform rash
3. Local lesion present
4. Louse transmission
5. Animal reservoir is wild house mouse

A. Rickettsialpox
B. Tsutsugamushi disease
C. Rocky Mountain spotted fever

Ref. 2 - pp. 839-840

SECTION III - BACTERIOLOGY

889.
1. Rickettsia tsutsugamushi
2. Rickettsia nipponica
3. Rickettsia orientalis
4. Rickettsia akamushi
5. Rickettsia akari

A. Synonyms for the same organism
B. Different etiologic agents for tsutsugamushi-like disease
C. Spread by Japanese beetles

Ref. 2 - p. 840

890.
1. Killed vaccines are ineffective in man
2. Gilliam and Karp strains commonly used as antigens
3. Effective immunity using live vaccine along with chloramphenicol
4. Formalized vaccine now in general use
5. Lack of cross-protection among various strains of the rickettsia

A. Tsutsugamushi disease
B. Rocky Mountain spotted fever
C. Rickettsialpox

Ref. 2 - p. 840

891.
1. Microtus montebilli (field mouse)
2. Mus concolor (house rat)
3. Mus diardii (field rat)
4. Mus musculus (wild house mouse)
5. Rattus flavipectus yunnanensis (wild rat)

A. Vertebrate host for rickettsialpox
B. Vertebrate host for Q fever
C. Vertebrate host for Rickettsia tsutsugamushi

Ref. 2 - p. 841

892.
1. Passes bacteria-proof filters
2. No toxin formation
3. Greater resistance to heat
4. Arthropod transmission not necessary
5. Cultivable on artificial media

A. Rickettsia prowazeki
B. Rickettsia akari
C. Coxiella burnetii

Ref. 2 - p. 841

893.
1. Described first in Australia
2. Incubation period of 14-26 days
3. Acquired by inhalation of infectious material
4. Atypical pneumonia
5. Macular rash

A. Pneumococcal pneumonia
B. Staphylococcal pneumonia
C. Q fever

Ref. 2 - p. 841

894.
1. Found in human excreta
2. Found in sheep
3. Found in milk
4. Found in cattle
5. Found in domestic fowl

A. Coxiella burnetii
B. Pathogenic spirochetes
C. Pathogenic amebae

Ref. 2 - p. 842

895.
1. Wolhynian fever
2. 6-22 days' incubation period
3. Pain in the legs
4. High fever
5. High mortality

A. French fever
B. Trench fever
C. Maladie de Roger

Ref. 2 - p. 842

896.
1. Infectious agent in the urine of patient
2. Infectious agent in louse feces
3. Infectious agent in blood of patient
4. Epidemics from unchlorinated swimming pools
5. Transmission by body louse

A. Borrelia recurrentis
B. Swimming pool conjunctivitis
C. Trench fever

Ref. 2 - p. 842

SECTION III - BACTERIOLOGY 103

897.
1. World-wide distribution
2. Vector : fleas
3. Vertebrate reservoir : dogs
4. Vector : rat louse
5. Vertebrate reservoir : wild rats

A. Rickettsia typhi
B. Rickettsia conori
C. Rickettsia akari

Ref. 2 - p. 832

898.
1. Permits routine identification of typhus
2. Permits routine identification of Brill's disease
3. Permits routine identification of scrub typhus
4. Permits routine identification of tsutsugamushi disease
5. Permits routine identification of spotted fever

A. Weil-Felix reaction
B. Toxin neutralization test
C. Protection of experimental animals against known challenge inoculum

Ref. 2 - p. 843

899.
1. European typhus
2. Scrub typhus
3. Classic typhus
4. Endemic typhus
5. Murine typhus

A. OX-19 ++++
B. OX-K +
C. Undetermined

Ref. 2 - p. 830

900.
1. Spotted fever
2. Indian tick typhus
3. São Paulo typhus
4. Kenya fever
5. Tsutsugamushi disease

A. OX-19 +
B. OX-K -
C. OX-2 -

Ref. 2 - p. 830

MATCH EACH OF THE FIVE ITEMS LISTED BELOW WITH THE MOST APPROPRIATE STATEMENTS OR CHOICES. EACH ITEM MAY BE USED ONLY ONCE:

A. Cardiolipin
B. Pretibial fever
C. Lister Institute
D. Vincent's angina
E. Frambesia

901. ___ Yaws
902. ___ Bacteroides fusiformis
903. ___ Wasserman antigen
904. ___ L. autumnalis
905. ___ L forms

Ref. 12 - pp. 890, 892, 885, 895, 912

A. R. akari
B. R. conori
C. R. australis
D. R. prowazekii
E. R. quintana

906. ___ Boutonneuse fever
907. ___ Trench fever
908. ___ Brill's disease
909. ___ Rickettsialpox
910. ___ Queensland tick typhus

Ref. 12 - p. 935

SECTION IV - VIROLOGY

FOR EACH OF THE FOLLOWING MULTIPLE CHOICE QUESTIONS, CHOOSE THE ONE MOST APPROPRIATE ANSWER:

911. MOST VIRUSES HAVE THE FOLLOWING COMPONENTS, EXCEPT:
   A. A centrally located nucleic acid
   B. Both RNA and DNA
   C. Either RNA or DNA
   D. A protein shell
   E. A nucleocapsid
   Ref. 14 - p. 39

912. ALL ENVELOPED HELICAL RNA VIRUSES BELONG TO ONE LARGE GROUP WHICH INCLUDES:
   A. Influenza
   B. Parainfluenza
   C. Measles
   D. Mumps
   E. All of the above
   Ref. 14 - p. 43

913. ORGANIZATION OF VIRUS PARTICLES:
   A. The viral DNA or RNA may be a single-stranded molecule
   B. The viral DNA or RNA may be a double-stranded molecule
   C. The chain may be linear
   D. The chain may be linked back on itself in a circle
   E. All of the above
   Ref. 4 - p. 53

914. THE FOLLOWING VIRUS IS COMPOSED OF TWO DISTINCT CAPSIDS ENCLOSING THE DOUBLE-STRANDED RNA:
   A. Herpesvirus
   B. Adenovirus
   C. Reovirus
   D. Poxvirus
   E. Myxovirus
   Ref. 14 - p. 46

915. MOST CASES OF POSTINFECTION ENCEPHALITIS FOLLOW:
   A. Rheumatic fever
   B. Chickenpox
   C. Mumps
   D. Measles
   E. Whooping cough
   Ref. 4 - p. 1177

916. THE INCUBATION PERIOD OF SERUM HEPATITIS RANGES FROM:
   A. 60-160 days
   B. 10-30 days
   C. 15-40 days
   D. 20-60 days
   E. 7-21 days
   Ref. 4 - p. 967

917. THE DIAMETER OF RABIES VIRUS HAS BEEN ESTIMATED BY FILTRATION TO BE:
   A. 100-150 mμ
   B. 70-75 mμ
   C. 60-70 mμ
   D. 40-60 mμ
   E. 20-40 mμ
   Ref. 4 - p. 816

918. THE DNA VIRUSES INCLUDE:
   A. Human adenoviruses
   B. Herpes simplex
   C. Varicella-zoster
   D. All of the above
   E. None of the above
   Ref. 4 - p. 14

919. THE MYXOVIRUS GROUP INCLUDES THE VIRUSES OF:
   A. Influenza, mumps, measles, Newcastle disease
   B. Herpes simplex, influenza, Newcastle disease, hepatitis
   C. Poliomyelitis, mumps, measles, echovirus
   D. Newcastle disease, influenza, smallpox, fowl plague
   E. None of the above
   Ref. 4 - p. 14

SECTION IV - VIROLOGY

920. THE FOUR PRINCIPLE WAYS OF MEASURING VIRUS-ANTIBODY REACTIONS ARE:
    A. Neutralization, electron microscopy, hemagglutination, precipitin
    B. Precipitin, neutralization, complement-fixation, hemagglutination-inhibition
    C. Ultracentrifugation, tissue culture, precipitin, hemagglutination-inhibition
    D. All of the above
    E. None of the above
    Ref. 4 - p. 224

921. DENGUE FEVER:
    A. The usual incubation period is from 2-3 weeks
    B. The fever may be sustained for several days or may have a diphasic course
    C. Absolute bradycardia is seen early
    D. The backache and pains in the muscles and the joints are mild
    E. Weakness and prostration are not characterisitc
    Ref. 4 - p. 618

922. THE PICORNAVIRUSES:
    A. Include numerous human enteroviruses of the poliovirus, coxsackie viruses A and B, and echovirus groups
    B. Are ether sensitive
    C. Hemagglutination of red cells is a property of all picornaviruses
    D. Are all stable at a pH as low as 3
    E. All of the above
    Ref. 4 - p. 95

923. ALL OF THE FOLLOWING STATEMENTS DESCRIBE THE HEMAGGLUTINATING INFLUENZA VIRUS, EXCEPT:
    A. The viral particle itself causes hemagglutination
    B. The active portion of the particle is known as the hemagglutinin
    C. The active portion is a superficially situated protein antigen
    D. The protein antigen also reacts in the complement fixation test
    E. The protein antigen does not react in the complement fixation test
    Ref. 14 - p. 59

924. AMONG THE FIRST ANIMAL VIRUSES TO BE SHOWN TO POSSESS THE CAPACITY TO CAUSE AGGLUTINATION OF RED BLOOD CELLS WAS:
    A. Foot and mouth disease virus
    B. Rubeola virus
    C. Encephalitis virus
    D. Poliomyelitis virus
    E. Influenza viruses
    Ref. 4 - pp. 693-694

925. COXSACKIE VIRUSES:
    A. Have been shown to be similar to polioviruses in size and sedimentation velocity
    B. Are associated with a characteristic cytopathic effect (CPE) in cultured cells like other enteroviruses
    C. Are relatively stable
    D. All of the above
    E. None of the above
    Ref. 4 - pp. 476, 477

926. NONSELECTIVE METABOLIC INHIBITORS OF VIRAL REPLICATION:
    A. Puromycin
    B. Actinomycin D
    C. P-fluorophenylalanine
    D. All of the above
    E. None of the above
    Ref. 4 - p. 305

SECTION IV - VIROLOGY

927. WHICH IS NOT A CHARACTERISTIC FEATURE OF THE ADENOVIRUSES?:
  A. The common unique fine structure of the virion (65-85 m diameter, 252 capsomeres)
  B. The presence of DNA
  C. The presence of RNA
  D. Ether-resistance
  E. The production of the "adenovirus type cytopathogenic effect"
  Ref. 4 - p. 151

928. WHEN THE MUMPS VIRUS AGGLUTINATES RED CELLS:
  A. There is a relatively firm adsorption of viral particle to mucopolysaccharide receptors on the surface of the red cell
  B. The adsorbed virus cannot be eluted even by warming the preparation to 37° C
  C. The same red cell is not agglutinable by other myxoviruses
  D. Hemagglutination cannot be inhibited specifically by antiserum
  E. All of the above
  Ref. 14 - p. 59

929. THE ONLY NATURAL HOST FOR MOST TYPES OF ADENOVIRUSES IS:
  A. Man
  B. Rat
  C. Mouse
  D. Guinea pig
  E. Monkey
  Ref. 4 - pp. 871, 875

930. EXANTHEM SUBITUM IS MOST COMMON IN THE FOLLOWING AGE GROUP:
  A. 20-30 years
  B. 7-10 years
  C. 13-19 years
  D. Few months to 2 years
  E. 5-15 years
  Ref. 4 - p. 811

931. THE MOST DEFINITIVE DISTINGUISHING FEATURE OF INFECTIOUS MONONUCLEOSIS FROM INFECTIOUS HEPATITIS:
  A. Fever
  B. Jaundice
  C. Heterophile antibodies
  D. Splenomegaly
  E. Leukocytosis
  Ref. 4 - p. 998

932. COXSACKIE VIRUSES OF GROUP B, TYPES 1 TO 6, ARE PROBABLY RESPONSIBLE FOR A CONSIDERABLE PROPORTION OF CASES OF:
  A. Poliomyelitis
  B. Roseola
  C. Mumps
  D. Aseptic meningitis
  E. None of the above
  Ref. 4 - p. 492

933. THE INCUBATION PERIOD FOR VARIOLA GENERALLY IS:
  A. 2-3 days
  B. 3-5 days
  C. 7 days
  D. 12 days
  E. 2-3 weeks
  Ref. 4 - pp. 938-939

934. THE PREGNANT WOMAN SEEMS TO FARE PARTICULARLY BADLY WHEN SHE CONTRACTS:
  A. Smallpox
  B. Rubella
  C. Mumps
  D. Influenza
  E. Scarlet fever
  Ref. 4 - p. 940

935. EPIDEMIC VIRAL GASTROENTERITIS AFFECTS THE FOLLOWING AGE GROUP:
  A. Few months to 2 years
  B. 5-10 years
  C. 20-30 years
  D. Older adults
  E. All ages
  Ref. 4 - p. 1171

SECTION IV - VIROLOGY

936. LESIONS OF MOLLUSCUM CONTAGIOSUM NEVER APPEAR ON:
    A. Arms and legs
    B. Buttocks
    C. Scalp and face
    D. Genitalia
    E. Soles and palms
    Ref. 4 - p. 1170

937. WHICH ONE IS NOT A DISEASE OF THE SKIN?:
    A. Herpes simplex
    B. Primary herpetic dermatitis
    C. Eczema herpeticum
    D. Acute herpetic rhinitis
    E. Traumatic herpes
    Ref. 4 - p. 900

938. THE FOLLOWING VIRUS CONTAINS STRUCTURALLY ESSENTIAL LIPID:
    A. Herpesvirus
    B. Papovavirus
    C. Adenovirus
    D. Reovirus
    E. Picornavirus
    Ref. 14 - p. 60

939. THE AREA MORE FREQUENTLY AFFECTED IN THE PARALYTIC FORM OF POLIOMYELITIS IN YOUNG CHILDREN:
    A. Arms
    B. Legs
    C. Bladder
    D. All of the above
    E. None of the above
    Ref. 4 - p. 447

940. DURING THE ERUPTIVE PHASE OF SMALLPOX, THE "MORE RECENT" LESIONS APPEAR ON THE:
    A. Lower limbs
    B. Face
    C. Arms
    D. Abdomen
    E. Chest
    Ref. 4 - p. 939

941. MOST FREQUENTLY INVOLVED IN HERPES ZOSTER:
    A. Dorsal roots of the trunk
    B. Sciatic nerves
    C. Facial nerves
    D. Perineal area
    E. Lower extremities
    Ref. 4 - p. 920

942. ON RECOVERING FROM INFECTION WITH HERPES VIRUS HOMINIS, THE PATIENT DEVELOPS:
    A. Delayed hypersensitivity
    B. Immediate hypersensitivity
    C. Arthus
    D. Mixed Arthus-delayed
    E. None of the above
    Ref. 4 - p. 896

943. THE FOLLOWING POXVIRUSES ARE SEROLOGICALLY RELATED, EXCEPT:
    A. Vaccinia
    B. Chickenpox
    C. Cowpox
    D. Rabbitpox
    E. Monkeypox
    Ref. 4 - p. 936

944. THE FOLLOWING RNA CARRIES FROM THE DNA GENOME IN THE NUCLEUS THE GENETIC INFORMATION, CODED IN ITS SEQUENCE OF BASES, TO BE TRANSLATED BY THE PROTEIN SYNTHESIZING SYSTEM IN THE CYTOPLASM OF THE CELL:
    A. rRNA
    B. mRNA
    C. sRNA
    D. All of the above
    E. None of the above
    Ref. 14 - p. 75

945. THE ONLY ANIMAL, OTHER THAN MAN, TO CONTRACT VARIOLA UNDER NATURAL CONDITIONS:
    A. Sheep
    B. Calf
    C. Monkey
    D. Goat
    E. Rabbit
    Ref. 4 - p. 935

SECTION IV - VIROLOGY

946. IN MUMPS, GLANDULAR SWELLING MOST OFTEN PERSISTS FOR:
    A. 1-2 days
    B. 3-5 days
    C. 5-7 days
    D. 7-10 days
    E. 14-18 days
    Ref. 4 - p. 759

947. THE FOLLOWING CHARACTERISTIC FINDING IS OF CONSIDERABLE DIAGNOSTIC AID IN POLIOMYELITIS:
    A. CSF sugar content
    B. High RBC
    C. High WBC
    D. Abnormal spinal fluid
    E. High blood pressure
    Ref. 4 - p. 449

948. THE PRODROMAL SYMPTOMS OF MEASLES ARE MOSTLY:
    A. Gastrointestinal disorders
    B. Central nervous system disorders
    C. Catarrhal
    D. Widespread rash
    E. Enlarged glands
    Ref. 4 - p. 792

949. IN CELLS INFECTED WITH VIRUS, ACTUAL SYNTHESIS OF VIRUS IS IMMEDIATELY PRECEDED BY:
    A. Attachment
    B. Penetration
    C. Eclipse
    D. Release
    E. None of the above
    Ref. 14 - p. 77

950. POLIOVIRUS VIRIONS WHICH BECOME ATTACHED TO HELA CELLS:
    A. Are not changed from their native state
    B. About 5 per cent of the attached virions usually elute at 37° C
    C. They do not bind antibody that has been induced by normal poliovirus
    D. They can still bind antibody that has been induced by normal poliovirus
    E. They are not susceptible to proteolytic degradation by trypsin
    Ref. 14 - p. 79

951. WHICH OF THE FOLLOWING STATEMENTS IS INCORRECT FOR ECHOVIRUSES?:
    A. They can be isolated only in cell cultures
    B. They produce varied diseases in laboratory animals
    C. They share many properties with rhinoviruses
    D. They are common causes of aseptic meningitis
    E. They have similar physical and chemical properties with the polioviruses and coxsackie viruses
    Ref. 4 - p. 513

952. THE GROUP B ARBOVIRUSES INCLUDE ALL OF THE FOLLOWING, EXCEPT:
    A. St. Louis encephalitis
    B. Japanese encephalitis
    C. Murray Valley encephalitis
    D. Dengue fever
    E. Venezuelan equine encephalitis
    Ref. 4 - p. 607

953. INFECTIOUS MONONUCLEOSIS MAY READILY BE CONFUSED WITH:
    A. Poliomyelitis
    B. Infectious hepatitis
    C. Influenza
    D. Mumps
    E. Encephalitis
    Ref. 4 - p. 998

954. THE MAIN MODE OF TRANSMISSION OF INFECTIOUS HEPATITIS:
    A. Person-to-person
    B. Water-borne
    C. Food-borne
    D. Milk-borne
    E. None of the above
    Ref. 4 - p. 979

SECTION IV - VIROLOGY   109

955. THE "DROMEDARY" COURSE OR 2-HUMPED TEMPERATURE CURVE IS REFERRED TO IN:
   A. Encephalitis
   B. Meningitis
   C. Poliomyelitis
   D. Scarlet fever
   E. Peritonitis
   Ref. 4 - p. 446

956. WHICH ONE OF THE FOLLOWING GROUPS OF VIRUSES POSSESSES AN ENZYME CAPABLE OF SPLITTING OFF NEURAMINIC ACID DERIVATIVES FROM SOME COMMON MUCOPROTEINS?:
   A. Pox viruses
   B. Herpes viruses
   C. Poliomyelitis virus
   D. Myxoviruses
   E. Mantle viruses
   Ref. 4 - p. 221

957. MYXOVIRUS HEMAGGLUTINATION:
   A. Is readily inhibited by specific antibody
   B. Inhibition cannot differentiate all the major types
   C. Inhibition cannot detect minor differences between strains of influenza A viruses
   D. Hemagglutination inhibition does not parallel neutralization titer
   E. Hemagglutination inhibition parallels the complement-fixation titer
   Ref. 4 - p. 220

958. VIRUSES ASSOCIATED WITH ACUTE RESPIRATORY ILLNESS INCLUDE:
   A. Rhinoviruses
   B. Adeno 1, 2, 3, 5, 14, 21
   C. Para-influenza 1, 3
   D. All of the above
   E. None of the above
   Ref. 4 - p. 415

959. IN THE SYNTHESIS OF POLIOVIRUS, THE POLIOVIRUS NUCLEIC ACID ACTS AS A mRNA TO PRODUCE PROTEIN WITHIN THE FOLLOWING PERIOD OF TIME AFTER THE CELL IS INFECTED:
   A. 30 minutes
   B. 3 hours
   C. 1 day
   D. 3 days
   E. None of the above
   Ref. 14 - p. 80

960. THE ARBOVIRUSES ARE DEFINED AS VIRUSES CAPABLE OF INFECTING CERTAIN:
   A. Vertebrates
   B. Mammals
   C. Mammals and vertebrates
   D. Vertebrates and multiplying in arthopods
   E. Birds
   Ref. 4 - p. 580

961. SEVERAL FAMILIES OF BLOOD-SUCKING ARTHROPODS ARE CLEARLY IMPLICATED AS NATURAL VECTORS OF ARBOVIRUSES. ONE OF THE FOLLOWING IS ONLY SUSPECTED:
   A. Culicidae
   B. Ceratopogonidae
   C. Lelaptidae
   D. Psychodidae
   E. Ixodidae
   Ref. 4 - p. 581

962. RABIES VIRUS IS RAPIDLY DESTROYED BY:
   A. Ultraviolet radiation
   B. Ether
   C. Phenol
   D. Sulfadiazine
   E. Thiomersal
   Ref. 4 - p. 817

963. THE SYMMETRIC PROTEIN SHELL WHICH ENCLOSES THE NUCLEIC ACID GENOME:
   A. Capsomere
   B. Capsid
   C. Virion
   D. Nucleocapsid
   E. None of the above
   Ref. 15 - p. 279

SECTION IV - VIROLOGY

964. THE MECHANISM BY WHICH SPECIFIC INFORMATION ENCODED IN A NUCLEIC ACID CHAIN IN A VIRUS IS TRANSFERRED TO MESSENGER RNA:
    A. Translation
    B. Transcription
    C. Transformation
    D. Transduction
    E. None of the above
    Ref. 15 - p. 279

965. THE PRIMARY CELLS IN WHICH ECHOVIRUSES MULTIPLY APPEAR TO BE LOCALIZED IN THE:
    A. Alimentary tract
    B. Central nervous system
    C. Respiratory system
    D. Reproductive system
    E. Blood and lymph systems
    Ref. 4 - p. 526

966. THE FOLLOWING VIRUS BELONGS TO THE RECENTLY PROPOSED RHABDOVIRUS GROUP THAT IS ROD-SHAPED, RESEMBLING A BULLET, FLAT AT ONE END AND ROUNDED AT THE OTHER:
    A. Rabies virus
    B. Yellow fever virus
    C. Vaccinia virus
    D. Colorado tick fever virus
    E. Rubella virus
    Ref. 15 - p. 284

967. VIRUSES WHICH ARE SUSCEPTIBLE TO THE ACTION OF SODIUM DEOXYCHOLATE:
    A. Coxsackie
    B. Mouse encephalomyelitis
    C. Encephalomyocarditis
    D. Poliomyelitis
    E. Arboviruses
    Ref. 4 - p. 580

968. THE CASE OF VARIOLA MAJOR MODIFIED BY PREVIOUS VACCINATION, OR ALASTRIM IN A VACCINATED OR UNVACCINATED PERSON, MAY BE CONFUSED MOST COMMONLY WITH ALL OF THE FOLLOWING, EXCEPT:
    A. Varicella
    B. Pustular acne
    C. Erythema multiforme
    D. Rocky Mountain spotted fever
    E. Drug eruptions
    Ref. 4 - p. 941

969. THE SYMPTOMS OF RUBELLA RESEMBLE THOSE OF MEASLES, WITH THE EXCEPTION OF:
    A. Fever
    B. Sneezing
    C. Koplik's spots
    D. Rash
    E. Running nose
    Ref. 4 - p. 806

970. AN ORGAN NOT EXHIBITING SIGNS OF MUMPS INVOLVEMENT:
    A. Eye
    B. Labyrinth
    C. Mammary glands
    D. Gall bladder
    E. Testis
    Ref. 4 - p. 760

971. IF THE PATIENT HAS FEVER, SORE THROAT AND LYMPHADENO-PATHY ACCOMPANIED BY LYMPHOCYTOSIS WITH ATYPICAL CELLS AND AN INCREASE IN SHEEP-CELL AGGLUTININS, THE DIAGNOSIS IS:
    A. Influenza
    B. Infectious mononucleosis
    C. Infectious hepatitis
    D. Toxoplasmosis
    E. Tuberculosis
    Ref. 4 - p. 997

972. THE FOLLOWING VIRUS MEASURES 225 x 300 mμ:
    A. Herpes simplex
    B. Rabies
    C. Mumps
    D. Vaccinia
    E. Influenza virus A
    Ref. 6 - p. 747

SECTION IV - VIROLOGY 111

973. THE INCUBATION PERIOD OF INFECTIOUS HEPATITIS RANGES FROM:
    A. 7 to 14 days
    B. 15 to 40 days
    C. 5 to 10 days
    D. 10 to 20 days
    E. 24 to 72 hours
    Ref. 4 - p. 967

974. ALL VIRUSES POSSESSING ENVELOPES CONTAIN COMPLEX MIXTURES OF:
    A. Neutral lipids
    B. Phospholipids
    C. Glycolipids
    D. All of the above
    E. None of the above
    Ref. 6 - p. 745

975. AEDES AEGYPTI IS THE VECTOR FOR:
    A. Louping ill
    B. Omsk hemorrhagic fever
    C. Yellow fever
    D. Kyasanur Forest disease
    E. Powassan
    Ref. 4 - p. 608

976. THE INCUBATION PERIOD OF INFLUENZA ORDINARILY IS:
    A. 4 to 6 days
    B. 2 to 4 days
    C. 1 to 2 days
    D. 3 to 5 days
    E. 5 to 7 days
    Ref. 4 - p. 704

977. THE MYXOVIRUS GROUP:
    A. Average particle size varies from about 100 mµ to 500 mµ
    B. All members of this group are RNA viruses
    C. Their nucleic acid is in the form of a nucleoprotein helix
    D. All of the above
    E. None of the above
    Ref. 4 - p. 685

978. MOST ANIMAL VIRUSES CONFORM TO:
    A. Two morphologic patterns
    B. Three morphologic patterns
    C. Five morphologic patterns
    D. Thirteen morphologic patterns
    E. None of the above
    Ref. 6 - p. 746

979. VIRAL PROTEIN IS ALWAYS SYNTHESIZED IN THE CYTOPLASM ON POLYRIBOSOMES COMPOSED OF:
    A. Viral messenger RNA
    B. Host cell ribosomes
    C. Host cell transfer RNA
    D. All of the above
    E. None of the above
    Ref. 6 - p. 772

980. IN INTERFERENCE BETWEEN VIRUSES:
    A. Generally, the first infecting virus is the one which interferes
    B. Sometimes, the first virus inhibits the ability of the second virus to adsorb
    C. Sometimes, the first virus prevents the early messenger RNA of the second virus from being translated
    D. All of the above
    E. None of the above
    Ref. 6 - p. 799

981. RABIES VIRUS HAS BEEN GROUPED WITH:
    A. Myxoviruses
    B. Adenoviruses
    C. Herpesviruses
    D. Pox viruses
    E. None of the above
    Ref. 6 - p. 752

982. INCREASED PRODUCTION OF ONE VIRUS AS THE RESULT OF CO-INFECTION WITH A SECOND VIRUS IS KNOWN AS:
    A. Interference
    B. Enhancement
    C. Complementation
    D. Phenotypic mixing
    E. Genotypic mixing
    Ref. 15 - p. 303

SECTION IV - VIROLOGY

983. THE CRITERIA FOR THE TREATMENT OF RABIES:
   A. The animal is apprehended and presents clinical signs of rabies
   B. The animal is killed and the brain is positive for Negri bodies
   C. The animal is killed and the brain is positive for rabies antigen as determined by the fluorescent antibody test
   D. A person is bitten by a stray animal that escapes
   E. All of the above   Ref. 4 - p. 827

984. FOOT-AND-MOUTH DISEASE IS A HIGHLY CONTAGIOUS DISEASE PRIMARILY OF THE FOLLOWING ANIMALS, EXCEPT:
   A. Cattle
   B. Swine
   C. Sheep
   D. Goats
   E. Man
   Ref. 2 - p. 930

985. THE LARGEST AND MOST COMPLEX OF ALL ANIMAL VIRUSES:
   A. Tumor viruses
   B. Arboviruses
   C. Picornaviruses
   D. Poxviruses
   E. Herpesviruses
   Ref. 6 - p. 731

986. VARICELLA:
   A. Is primarily a disease of children
   B. Has an incubation period of less than 14 days
   C. The rash occurs all at once
   D. The rash occurs first at the extremities
   E. The lesion is usually hemorrhagic   Ref. 6 - p. 864

987. A SPECIAL FEATURE THAT CHARACTERIZES THE VACCINIA VIRUS:
   A. Single-stranded RNA
   B. Brick-shaped
   C. Naked DNA
   D. Bullet-shaped
   E. Double-stranded RNA
   Ref. 12 - p. 1016

988. THE DNA OF EVEN-NUMBER COLIPHAGES IS CONTAINED IN THE:
   A. Head
   B. Collar
   C. Sheath
   D. Plate
   E. Fibers
   Ref. 12 - p. 1034

989. ALL OF THE FOLLOWING TYPES OF NUCLEIC ACID CAN OCCUR IN VIRUSES, EXCEPT:
   A. Single-stranded DNA
   B. Double-stranded DNA
   C. Single-stranded RNA
   D. Double-stranded RNA
   E. Both DNA and RNA
   Ref. 12 - p. 1036

990. THE TISSUE ORIGIN OF HELA CELLS IS:
   A. Sternal bone marrow
   B. Embryonic lung
   C. Carcinoma, cervix
   D. Connective tissue
   E. Kidney
   Ref. 12 - p. 1135

991. THE DNA OF ALL ANIMAL VIRUSES IS SYNTHESIZED IN THE CELL NUCLEUS, EXCEPT FOR THE FOLLOWING GROUP OF VIRUSES:
   A. Herpesviruses
   B. Adenoviruses
   C. Poxviruses
   D. Picornaviruses
   E. Myxoviruses
   Ref. 12 - p. 1141

992. LONG-LASTING IMMUNITY RESULTS FROM THE FOLLOWING DISEASES, EXCEPT:
   A. Smallpox
   B. Measles
   C. Yellow fever
   D. Influenza
   E. Poliomyelitis
   Ref. 12 - p. 1210

SECTION IV - VIROLOGY

993. THE INCUBATION PERIOD OF VARICELLA IS:
   A. 5 to 7 days
   B. 7 to 10 days
   C. 14 to 16 days
   D. 18 to 21 days
   E. None of the above
   Ref. 12 - p. 1245

994. POXVIRUSES ARE CHARACTERIZED BY THE FOLLOWING, EXCEPT:
   A. They are brick-shaped to ovoid
   B. They are relatively resistant to chemical and physical inactivation
   C. They contain double-stranded RNA
   D. They have a predilection for epidermal cells
   E. They produce eosinophilic cytoplasmic inclusion bodies
   Ref. 12 - p. 1255

995. COXSACKIE VIRUSES ARE SIMILAR TO POLIOVIRUSES IN THE FOLLOWING CHARACTERISTICS, EXCEPT:
   A. Size
   B. Molecular weight
   C. Base composition of the RNA
   D. Shape
   E. Stability
   Ref. 12 - p. 1296

996. ORCHITIS OCCURS AS A CONSEQUENCE OF MUMPS IN THE FOLLOWING PERCENTAGE OF MALES PAST PUBERTY:
   A. 5 to 10 percent
   B. 10 to 20 percent
   C. 25 to 35 percent
   D. 50 to 75 percent
   E. Almost 100 percent
   Ref. 12 - p. 1355

997. LIKE MOST MYXOVIRUSES AND PARAMYXOVIRUSES, MEASLES VIRUS:
   A. Agglutinates monkey RBC best at 37^0
   B. Elutes spontaneously from agglutinated cells
   C. Agglutinates red blood cells and virus-specific antibodies inhibit the hemagglutination
   D. The erythrocyte receptors are destroyed by neuraminidase
   E. None of the above
   Ref. 12 - p. 1357

998. IN SERUM HEPATITIS:
   A. The virus can be detected in duodenal contents and feces
   B. The virus is usually present in the blood during incubation period and acute phase
   C. Fever over 38^0 is common
   D. The incubation period is about 50 days
   E. The icteric stage has an abrupt onset
   Ref. 12 - p. 1409

999. CONCERNING THE HERPES SIMPLEX VIRUS:
   A. The DNA is single-stranded
   B. The DNA is double-stranded
   C. It is stable at room temperature
   D. It is not readily inactivated by lipid solvents
   E. Viral infectivity is best at acid pH
   Ref. 12 - p. 1240

1000. IN INFECTIOUS HEPATITIS:
   A. The onset may be insidious
   B. The fever may be low
   C. Viremia is present
   D. The virus is present in stools
   E. All of the above
   Ref. 6 - pp. 936-937

SECTION IV - VIROLOGY

ANSWER THE FOLLOWING QUESTIONS BY USING THE KEY OUTLINED BELOW:
- A. If both statement and reason are true and related cause and effect
- B. If both statement and reason are true but not related cause and effect
- C. If the statement is true but the reason is false
- D. If the statement is false but the reason is true
- E. If both statement and reason are false

1001. The clinical diagnosis of rubella may be difficult in an epidemic period BECAUSE classically it appears in the late winter and spring months in temperate zones. Ref. 4 - p. 806

1002. Rabies vaccine should not be given unless there is good evidence of exposure to rabies BECAUSE sensitization to rabbit-brain tissue may produce serious allergic reactions. Ref. 4 - p. 827

1003. Animal viruses with helical structure are usually more easy to purify BECAUSE of their lability and an envelope which contains lipid components derived from the host cell. Ref. 14 - p. 49

1004. It is suggested that active immunity induced by mumps infection cannot be passively transferred from mother to offspring via the placenta BECAUSE mumps is not rare in children under 6 or even 9 months of age. Ref. 4 - p. 758

1005. A useful distinction can be made between DNA- and RNA-containing viruses by the use of 5-fluorodeoxyuridine BECAUSE the growth of RNA viruses is not inhibited by this substance. Ref. 14 - p. 53

1006. The determination of the sequence of nucleotides in a molecule of the RNA of picornavirus is easy BECAUSE the picornavirus is small (25 mμ) with a molecular weight of 2×10^6. Ref. 14 - p. 55

1007. Definitive diagnosis of lymphocytic choriomeningitis rests on isolation and identification of the agent BECAUSE the clinical course cannot be differentiated from that caused by infection with a number of other neurotropic viruses. Ref. 4 - p. 1168

1008. The central feature of the eclipse phase of viral replication is the uncoating phenomenon BECAUSE in order for viral synthesis to occur, the protein coat surrounding the viral nucleic acid must be opened and the nucleic acid released "naked" into the cell. Ref. 14 - p. 78

1009. If warts are made to bleed, new lesions may occur in neighboring parts of the skin BECAUSE the virus cells are circulating in the blood stream. Ref. 4 - pp. 850-851

1010. Fatalities are rare and confined to aged persons stricken with epidemic viral gastroenteritis BECAUSE the extremely high temperature and respiratory difficulty accompanying the virus are poorly tolerated in this age group. Ref. 4 - p. 1171

1011. Lymphocytic choriomeningitis is an endemic viral infection of higher animals and of major importance BECAUSE infection by this virus may assume different clinical forms such as aseptic meningitis, grippe and acute fatal systemic disease. Ref. 4 - pp. 1166, 1168

SECTION IV - VIROLOGY

1012. Fatal cases of cat-scratch fever are not at all rare BECAUSE the mere licking of a cat can cause the disease.   Ref. 4 - p. 1166

1013. The presence of specific 19S antibody globulin to measles virus is an indication of recent primary infection or artificial antigenic stimulation BECAUSE 7S globulin does not appear until 21 days after infection.
Ref. 4 - p. 790

1014. Bed rest is most advisable for patients in the acute stage of infectious mononucleosis BECAUSE there is always the danger of rupture of the spleen.   Ref. 4 - p. 1000

1015. Infectious mononucleosis is rarely fatal BECAUSE broad-spectrum antibiotics are helpful in preventing complications.
Ref. 4 - pp. 1000-1001

1016. The control of varicella by isolation and quarantine is difficult BECAUSE of the extremely high order of infectivity of the patient in the initial stages of the disease.   Ref. 4 - p. 922

1017. Transplacental immunity to varicella in the newborn is absolute BECAUSE it is of the same order as in measles.   Ref. 4 - p. 923

1018. Antiserum against the virus will neutralize a virus BECUASE it reacts with the antigens of the protein coat, but the same antiserum is without effect when in contact with the free infectious viral nucleic acid.
Ref. 15 - p. 279

1019. The structural proteins of the viruses are not important in vaccine production or in viral diagnosis BECAUSE the proteins do not necessarily determine the antigenicity of the viruses.
Ref. 15 - p. 287

1020. Keratitis with impairment of vision may persist in from 1 to 10% of the patients afflicted with conjunctivitis due to adenovirus BECAUSE improper care and management can lead to ulceration of the cornea.
Ref. 4 - p. 878

1021. Adenovirus cannot be recovered from stools of ill persons BECAUSE gastrointestinal symptoms and signs are not a feature of this disease.
Ref. 4 - pp. 878, 879

1022. Ether susceptibility has been useful for distinguishing viruses that possess a lipid-rich envelop from those that do not BECAUSE the herpesviruses are inactivated by ether while the picornaviruses are resistant to ether.   Ref. 15 - p. 293

1023. Sulfonamides and antibiotics do not affect the influenza virus but should be administered as a prophylactic measure BECAUSE they are believed to prevent complications.   Ref. 4 - p. 710

1024. Virus multiplication was first studied successfully in the T-series of bacteriophages BECAUSE of their short growth cycle and their relative ease of quantification.   Ref. 15 - p. 295

1025. Cowpox is not a common disease BECAUSE it is rarely spread from person to person.   Ref. 4 - p. 953

1026. There is strong evidence that smallpox is infectious during the incubation period BECAUSE during this period there are lesions in the mouth and the upper respiratory tract.   Ref. 4 - p. 936

SECTION IV - VIROLOGY

1027. Inclusion bodies may be of considerable diagnostic aid BECAUSE its presence is pathognomonic for certain diseases.
Ref. 15 - p. 310

1028. Lymphocytic choriomeningitis is a rarely fatal malady BECAUSE the virus is susceptible to penicillin.   Ref. 4 - pp. 1167, 1169

1029. The degree of communicability of mumps is very low BECAUSE the infection occurs chiefly in children and adolescents.
Ref. 4 - p. 763

1030. Transmission of measles to susceptible contacts usually occurs before a diagnosis is established BECAUSE the major period of virus dessemination precedes the rash.   Ref. 4 - p. 795

1031. In addition to their medical importance, viruses provide the simplest model systems for many basic problems in biology BECAUSE viruses are essentially segments of genetic material encased in protective shells.   Ref. 6 - p. 714

1032. Plaque formation is often the most desirable method of titrating viruses BECAUSE all viruses can be measured in this way.
Ref. 6 - p. 720

1033. Infectious mononculeosis is common among medical personnel BECAUSE the disease is transmitted easily from person to person.
Ref. 4 - p. 1003

1034. Infectious mononculeosis is easily confused with infectious hepatitis BECAUSE during the early stages the blood picture is very similar.
Ref. 4 - p. 998

1035. Although mumps usually confers a durable immunity, it can recur BECAUSE when the disease attacks only a single parotid gland, the other parotid still remains susceptible to the disease.
Ref. 4 - pp. 758, 759

1036. Mumps is more common in torrid climates than in temperate climates BECAUSE there is evidence it can be transmitted by insects found in warmer climates.   Ref. 4 - p. 763

1037. The binomial <u>Herpesvirus varicellae</u> has been proposed BECAUSE the varicella-zoster (V-Z) virus is similar in many respects to the herpes simplex virus.   Ref. 4 - p. 916

1038. V-Z virus may be propagated in a variety of primary cultures of human tissues BECAUSE the virus produces a focal cytopathic process.
Ref. 4 - p. 917

1039. Quarantine in cases of chickenpox is usually of no benefit BECAUSE the disease is most communicable prior to the manifestation of symptoms.
Ref. 4 - p. 922

1040. Cases of shingles are rarely seen in patients under 20 years of age BECAUSE the disease does not appear in epidemics.
Ref. 4 - p. 922

1041. Rubella can be distinguished from measles BECAUSE rubella does not usually present so severe a clinical picture and displays a less extensive rash of shorter duration.   Ref. 4 - p. 793

SECTION IV - VIROLOGY

1042. Newborn infants remain protected against measles for approximately the first 6 to 9 months of life BECAUSE of transplacentally-acquired antibody.
Ref. 4 - p. 795

1043. The most common indirect method of measuring the number of virus particle is the hemagglutination assay BECAUSE all animal viruses adsorb to the red blood cells of various animal species.
Ref. 6 - p. 722

1044. The virus depends on host cells for most, if not all, of the small-molecular precursors used in viral synthesis BECAUSE the host cell supplies the energy required in viral synthesis.
Ref. 4 - p. 305

1045. The circularity of viral nucleic acids is essential for infectivity BECAUSE circularity is also essential for oncogenicity.
Ref. 6 - p. 740

1046. Pathogenicity is a consequence of the perturbation of the host's metabolism which is produced by viral nucleic acid BECAUSE the essential function of a virus particle is to transmit the infectious nucleic acid to a susceptible host.
Ref. 4 - p. 51

1047. The complete amino acid sequence of the protein subunit is known for the tobacco mosaic virus (TMV) BECAUSE it is the easiest virus to obtain in quantity in purified form.
Ref. 4 - p. 67

1048. It is not merely a matter of labeling viruses as DNA- or RNA-containing, but also of distinguishing them in terms of the amount of information carried by the nucleic acid BECAUSE the amount of information transmitted by a virus would depend on the size of its nucleic acid component.
Ref. 4 - p. 90

1049. Animal viruses in general attach to animal cells with far greater specificity than is seen with bacteriophages BECAUSE they attach and infect more efficiently than bacterial viruses.
Ref. 4 - p. 217

1050. One of the outstanding biologic properties of interferon is the virus specificity BECAUSE only a few viruses have been shown to be sensitive to interferon.
Ref. 4 - p. 328

1051. Viral RNAs cannot be readily separated from cellular RNAs BECAUSE of their large size and high sedimentation constant.
Ref. 12 - p. 1041

1052. Temperate phages produce turbid plaques BECAUSE they lyse only a fraction of the sensitive cells they infect.
Ref. 12 - p. 1101

1053. Interferons are very easy to obtain in pure form BECAUSE they are produced in large amounts.
Ref. 12 - p. 1171

1054. Dissemination of the poliovirus from individuals vaccinated with the live attenuated virus to unvaccinated individuals may possibly be a potential hazard BECAUSE if the transmission is uncontrolled, the viruses may change in virulence.
Ref. 12 - p. 1294

1055. Echoviruses are no longer "orphans" in reference to human diseases BECAUSE most echoviruses have been associated with one or more clinical syndromes.
Ref. 12 - p. 1303

SECTION IV - VIROLOGY

1056. Although RNA was extracted from only type 11 of rhinoviruses, others are considered to be RNA viruses BECAUSE they can multiply in the presence of 5-fluoro-deoxyuridine (FUdR), which inhibits propagation of DNA-containing viruses. Ref. 12 - p. 1306

1057. Definitive diagnosis of rabies in man, and in suspected animals, depends upon identification of Negri bodies in brain tissue BECAUSE they are most abundant in Ammon's horn of the hippocampus.
Ref. 12 - p. 1338

1058. Mumps can be prevented by immunization with live attenuated virus vaccine BECAUSE detectable antibodies are produced in 95% of vaccines.
Ref. 15 - p. 402

1059. Epidemic spread of measles is rapid BECAUSE the virus is shed early before the disease can be recognized. Ref. 12 - p. 1359

1060. Arboviruses often cause disease in both vertebrates and arthropods BECAUSE they multiply in both vertebrates and arthropods.
Ref. 12 - p. 1376

ANSWER THE FOLLOWING QUESTIONS BY USING THE KEY OUTLINED BELOW:
A. If only A is correct
B. If only B is correct
C. If both A and B are correct
D. If neither A nor B is correct

1061. DEPENDING ON THE TYPE OF VIRUS, THE TYPE OF CELLS AND CIRCUMSTANTIAL CONDITIONS, THE DNA VIRUSES IN A CELL CAN ENTER:
A. A dependent state
B. An independent state
C. Both
D. Neither
Ref. 4 - p. 267

1062. LABORATORY PROCEDURES AVAILABLE TO ESTABLISH DIAGNOSIS OF YELLOW FEVER INCLUDE:
A. Demonstration of the development of specific antibodies during infection
B. Pathologic examination of the liver in fatal cases
C. Both
D. Neither
Ref. 4 - p. 612

1063. OTHER NAMES FOR DENGUE ARE:
A. Break-bone fever
B. Eland fever
C. Both
D. Neither
Ref. 4 - p. 615

1064. THE MAJOR GROUPS OF ANIMAL VIRUSES ARE DEFINED IN TERMS OF:
A. Nucleic acid composition
B. Size
C. Both
D. Neither
Ref. 4 - p. 11

1065. THE INFLUENZA EPIDEMIC OF 1957 WAS CAUSED BY:
A. Influenza A virus (PR8)
B. Asian A_2
C. Both
D. Neither
Ref. 4 - p. 711

SECTION IV - VIROLOGY

1066. THE LIVE ATTENUATED RUBELLA VIRUS VACCINE IS RECOMMENDED FOR:
A. Postpubertal female
B. Those sensitive to chicken protein
C. Both
D. Neither
Ref. 15 - p. 411

1067. VIRIONS (MATURE VIRUS PARTICLES) ARE DIVIDED INTO TWO CATEGORIES:
A. Those which consist of a nucleocapsid surrounded by an envelope
B. Those which do not possess an envelope
C. Both
D. Neither
Ref. 4 - p. 12

1068. THE NAME, PICORNAVIRUS, IS DERIVED FROM:
A. Location in which it was first found
B. The fact that they contain RNA
C. Both
D. Neither
Ref. 4 - p. 13

1069. REOVIRUSES ARE UNIQUE IN THAT:
A. The RNA is double-stranded
B. The RNA is single-stranded
C. Both
D. Neither
Ref. 4 - p. 13

1070. "COLD" POLIOVIRUS MUTANTS CAN GROW ONLY BETWEEN ABOUT 24° AND 34° C, WHILE "HOT" STRAINS CAN GROW BETWEEN ABOUT 31° AND 41° C. THE MORE VIRULENT STRAINS ARE THOSE THAT CAN GROW IN THE:
A. Lower range
B. Higher range
C. Both
D. Neither
Ref. 6 - p. 795

1071. PAPOVAVIRUS GROUP:
A. Includes human papilloma (wart)
B. Refers to a group of small DNA tumor-inducing viruses
C. Both
D. Neither
Ref. 4 - p. 15

1072. THE MEMBERS OF THE POXVIRUS GROUP HAVE THE FOLLOWING PROPERTIES:
A. They are large brick-shaped viruses
B. They are all ether-sensitive
C. Both
D. Neither
Ref. 4 - p. 16

1073. THE MYXOVIRUSES:
A. Are the smallest of the RNA viruses
B. Were so named because of their affinity for mucoproteins
C. Both
D. Neither
Ref. 4 - p. 119

1074. RECENT INFORMATION ON THE BIOLOGIC, CHEMICAL, AND PHYSICAL PROPERTIES OF THE DNA VIRUSES HAS MADE IT POSSIBLE TO CLASSIFY THEM INTO:
A. 2 major groups
B. 10 major groups
C. Both
D. Neither
Ref. 15 - p. 281

SECTION IV - VIROLOGY

1075. THE SWITCH-OFF PHENOMENON REPRESENTING CESSATION OF THE SYNTHESIS OF EARLY ENZYMES OCCURS DURING:
A. The early period of vaccinia virus multiplication cycle
B. Commencement of the late period of vaccinia virus multiplication cycle
C. Both
D. Neither
Ref. 6 - p. 777

1076. ONE OF THE MAJOR CRITERIA FOR EXCLUDING THE PSITTACOSIS-LYMPHOGRANULOMA GROUP OF AGENTS AS "TRUE" VIRUSES IS:
A. The presence of both DNA and RNA
B. The size of DNA present
C. Both
D. Neither
Ref. 4 - p. 179

1077. THE MEMBERS OF THE HERPESVIRUS GROUP:
A. Are medium-sized DNA viruses
B. Possess an envelope
C. Both
D. Neither
Ref. 4 - p. 161

1078. MANY VIRULENT E. COLI PHAGES IN MULTIPLYING:
A. Always lead to death of the cell they infect
B. Do not lead to death of the cell they infect
C. Both
D. Neither
Ref. 4 - p. 183

1079. WHEN A BACTERIOPHAGE IS GROWN IN ONE HOST AND THEN LYSOGENIZES ANOTHER HOST:
A. It may transfer certain genetic characteristics from the first to the second
B. It cannot transfer genetic characteristics from the first to the second
C. Both
D. Neither
Ref. 4 - p. 301

1080. WHEN T2 ENTERS A BACTERIAL CELL:
A. That cell is immediately killed
B. The synthesis of host-type DNA and RNA stops at once
C. Both
D. Neither
Ref. 4 - p. 299

1081. THE MYXOVIRUSES HAVE AN OUTER ENVELOPE THAT CONTAINS:
A. A hemagglutinin
B. Neuraminidase
C. Both
D. Neither
Ref. 4 - p. 294

1082. SPONTANEOUS MUTATIONS APPEAR:
A. Only during the course of intracellular virus replication
B. Also during the extracellular resting stage
C. Both
D. Neither
Ref. 4 - p. 289

1083. BEST EVIDENCE FOR THE OCCURRENCE OF LATENT INFECTION WITH VIRUSES IN MAN:
A. Recurrent herpes simplex
B. Varicella-zoster virus
C. Both
D. Neither
Ref. 4 - p. 342

SECTION IV - VIROLOGY

1084. PRIMARY INFECTION WITH HERPES SIMPLEX IS USUALLY ESTABLISHED DURING INFANCY:
   A. As a recognizable gingivostomatitis
   B. As an inapparent infection to which circulating antibodies and hypersensitivity develop
   C. Both
   D. Neither   Ref. 4 - p. 342

1085. SOME VIRUSES INDUCE INCLUSION BODIES ONLY WITHIN CYTOPLASM, SUCH AS:
   A. Guarnieri bodies of vaccinia   C. Both
   B. Bollinger bodies of fowlpox   D. Neither
                                        Ref. 4 - p. 345

ANSWER THE FOLLOWING QUESTIONS (T)RUE OR (F)ALSE:

1086. The sequence of nucleotides determines the genetic message carried by the viral nucleic acids and directs the order of amino acids in protein, thereby producing biologic and antigenic specificity of the virus.
                                        Ref. 14 - p. 55

1087. Most of the work on the construction of elaborate linkage maps of the position of genes on certain viral chromosomes has been accomplished with bacteriophage, rather than with animal viruses.
                                        Ref. 14 - p. 56

1088. Viruses capable of causing transduction do not carry small segments of host nucleic acid in addition to that of the viral genome.
                                        Ref. 14 - p. 56

1089. Transcapsidation refers to the phenomenon which involves a helper virus, e.g., adenovirus which serves as a helper in transcapsidation of SV40 DNA.   Ref. 14 - p. 57

1090. The basic mechanisms by which a virus is replicated in an infected cell are the same as those employed in a normal cell to synthesize nucleic acids and protein.   Ref. 14 - p. 75

1091. Following the eclipse phase, the next step essential for viral synthesis is the programming of the infected cell by viral nucleic acid.
                                        Ref. 14 - p. 79

1092. The mechanism of viral release is the same among the viral groups although the sites of maturation differ widely.
                                        Ref. 14 - p. 81

1093. Within a single cycle of viral infection in culture, the tinctorial affinities of the inclusion never change, e.g., from basophilic to eosinophilic.
                                        Ref. 14 - pp. 85-87

1094. In cell culture, following attachment of the virion by its outer coat to the surface of the host cell, it is taken into the cell, intact, by a process of phagocytic engulfment known as viropexis and is enclosed in a vesicle.   Ref. 14 - p. 88

1095. Virion is the complete infective virus particle, which may be identical with the nucleocapsid.   Ref. 15 - p. 279

SECTION IV - VIROLOGY

1096. The capsids of animal viruses are arranged in helical symmetry but not in cubical symmetry.   Ref. 15 - p. 286

1097. In the replication of poliovirus, the single-stranded RNA can serve as its own messenger RNA.   Ref. 15 - p. 296

1098. In the replication of the DNA viruses, host-cell DNA synthesis is temporarily elevated and is then suppressed as the cell shifts over to the manufacture of viral DNA.   Ref. 15 - p. 298

1099. Recent evidence indicates that many agents that are potent interferon stimulators have in common the presence of double-stranded RNA in the material inoculated, or the production of double-stranded RNA during the replication of the virus.   Ref. 15 - p. 307

1100. In the implantation and multiplication of polioviruses in the host, spread to the CNS is not interrupted by the presence of antibodies, induced by prior infection or vaccine.   Ref. 15 - p. 308

1101. Infectious viral nucleic acid can be extracted not only from infectious virus particles, but from any virus particles which contain undamaged nucleic acid.   Ref. 6 - p. 743

1102. Formaldehyde destroys infectivity of viruses primarily by reacting with those amino groups of adenine, guanine, and cytosine which are not involved in hydrogen-bond formation.   Ref. 6 - p. 758

1103. Viruses containing double-stranded nucleic acid are inactivated readily by formaldehyde, while those which contain single-stranded nucleic acid are resistant to it.   Ref. 6 - p. 758

1104. In a one-step growth cycle of virus, the interval between the beginning of the late and rise periods represents the time necessary for the incorporation of a viral genome into a mature virion.
   Ref. 6 - pp. 768-769

1105. The location where the viral genome replicates is identical for all viruses.   Ref. 6 - p. 772

ANSWER THE FOLLOWING QUESTIONS BY USING THE KEY OUTLINED BELOW:
A. If A is greater in frequency and/or magnitude than B
B. If B is greater in frequency and/or magnitude than A
C. If A and B are approximately equal

1106. DISTRIBUTION OF POLIOMYELITIS IN AGE GROUPS:
A. Young children
B. Older individuals   Ref. 4 - p. 451

1107. IN THE DIAGNOSIS OF POLIOVIRUS INFECTIONS:
A. Neutralization test
B. Complement fixation test   Ref. 4 - p. 449

1108. PORTAL OF ENTRY OF POLIOMYELITIS VIRUS:
A. Olfactory route
B. Upper alimentary tract   Ref. 4 - p. 438

SECTION IV - VIROLOGY

1109. THE OCCURRENCE OF MEASLES IN EPIDEMIC FORM:
A. 5-year intervals
B. 3-year intervals
Ref. 4 - p. 795

1110. TREATMENT FOR COMMON COLD:
A. Antihistamine drugs
B. Symptomatic
Ref. 4 - p. 553

1111. FOOT-AND-MOUTH DISEASE:
A. In cattle
B. In man
Ref. 4 - p. 559

1112. THE OBSERVANCE OF FEVER OVER 38° C:
A. Infectious hepatitis
B. Serum hepatitis
Ref. 4 - p. 967

1113. DEMONSTRATION OF VIRUS IN FECES:
A. Serum hepatitis
B. Infectious hepatitis
Ref. 4 - p. 967

1114. THE AVERAGE INCUBATION PERIOD OF:
A. Infectious hepatitis
B. Serum hepatitis
Ref. 6 - p. 937

1115. PATHOLOGIC CHANGES PRODUCED BY:
A. Varicella
B. Herpes zoster
Ref. 15 - p. 435

1116. PEAK INCIDENCE OF INFECTIOUS MONONUCLEOSIS:
A. In young adults (21 to 25)
B. In older adults (31 to 35)
Ref. 4 - p. 1002

1117. DURATION OF IMMUNITY ACQUIRED IN DOGS VACCINATED WITH:
A. Flury LEP vaccine
B. Semple type vaccine
Ref. 4 - p. 833

1118. PEAK AGE OF MEASLES IN U.S.:
A. Between 1 and 3 years
B. Between 3 and 5 years
Ref. 4 - p. 795

1119. THE GLANDS INVOLVED IN MUMPS:
A. Submaxillary gland
B. Parotid gland
Ref. 4 - p. 759

1120. PROTECTION OF CNS FROM SUBSEQUENT INVASION BY WILD VIRUS:
A. Killed poliovirus vaccine
B. Live attenuated poliovirus vaccine
Ref. 15 - pp. 364-365

1121. THE MOST EFFECTIVE PREVENTIVE MEASURE OF MEASLES:
A. Live attenuated measles virus vaccine
B. Commercial gamma globulin
Ref. 15 - p. 405

1122. THE ETIOLOGY OF UPPER RESPIRATORY INFECTION (URI):
A. Para-influenza 1, 3
B. Para-influenza 2, 4
Ref. 4 - p. 415

1123. VIRAL DISEASES MAINTAINED BY DIRECT TRANSFER FROM MAN TO MAN AND FOR WHICH THERE IS NO KNOWN HOST, OTHER THAN MAN, ORDINARILY ARE SPREAD BY:
A. Droplet inhalation
B. Fecal contamination
Ref. 4 - p. 388

SECTION IV - VIROLOGY

1124. THE DISEASES IN WHICH THE CAUSATIVE VIRUSES PERSIST IN THE TISSUE OF THE RECOVERED HOSTS:
A. Herpes simplex
B. Infectious hepatitis   Ref. 4 - p. 389

1125. THE ARTHROPOD-TRANSMITTED VIRAL DISEASE IN WHICH THE SOURCE IS A HUMAN CASE:
A. Yellow fever
B. Dengue   Ref. 4 - p. 391

1126. TRANSMISSION OF SERUM HEPATITIS VIRUS TO MAN:
A. By administration of blood from individuals who harbor the virus
B. By instruments contaminated with the virus
                                           Ref. 4 - p. 393

1127. IN GENERAL, SEROLOGIC TESTS ARE USED IN DIAGNOSTIC VIROLOGY:
A. To identify the virus by testing it against a number of specific antisera
B. To demonstrate a rise in antibody against his own or the prototype virus strain   Ref. 4 - p. 409

1128. THE ETIOLOGY OF ORCHITIS AND EPIDIDYMITIS:
A. Mumps
B. Coxsackie B   Ref. 4 - p. 410

1129. ADVANTAGE OF LIVE VIRUS VACCINE OVER INACTIVATED VACCINE:
A. The speed with which the immune response develops
B. The persistence of immunity   Ref. 4 - p. 458

1130. TRANSMISSION OF ARBOVIRUSES, GROUP B:
A. Mosquitoes
B. Ticks   Ref. 4 - p. 607

MATCH EACH OF THE LETTERED ITEMS WITH THE MOST APPROPRIATE STATEMENTS OR CHOICES. EACH ITEM MAY BE USED ONLY ONCE:

A. Poxvirus group
B. Picornavirus group
C. Myxovirus group
D. Papovavirus group
E. Herpesvirus group

1131. ___ Varicella-zoster
1132. ___ Echoviruses
1133. ___ Wart
1134. ___ Ectromelia
1135. ___ Mumps   Ref. 4 - pp. 14-15

A. Enteritis (diarrhea)
B. Encephalitis
C. Pleurodynia
D. Orchitis
E. Herpangina

1136. ___ Coxsackie B1-6
1137. ___ Coxsackie A2, 4-6, 8, 10
1138. ___ Echo 11, 14, 17, 18
1139. ___ Arboviruses
1140. ___ Mumps   Ref. 4 - p. 410

SECTION IV - VIROLOGY

A. Influenza A
B. Asian
C. Influenza B
D. A-prime
E. Swine influenza

1141. ___ Prototype Lee/40
1142. ___ Prototype SW 15
1143. ___ Prototype Japan
1144. ___ Prototype PR8-WS
1145. ___ Prototype FM_1

Ref. 4 - p. 699

A. Mumps
B. Rabies
C. St. Louis encephalitis
D. Measles
E. Venezuelan equine encephalitis (VEE)

1146. ___ Rhabdovirus
1147. ___ Arbovirus Group A
1148. ___ Pseudomyxovirus
1149. ___ Paramyxovirus
1150. ___ Arbovirus Group B

Ref. 6 - pp. 750-752

A. Poxvirus group
B. Picornavirus group
C. Myxovirus group
D. Adenovirus group
E. Herpesvirus group

1151. ___ RNA virus, 17 - 30 mμ, ether resistant
1152. ___ RNA virus, 80 - 200 mμ, ether sensitive
1153. ___ DNA virus, 150 - 300 mμ, ether sensitive or resistant
1154. ___ DNA virus, 120 - 180 mμ, ether sensitive
1155. ___ DNA virus, 65 - 85 mμ, ether resistant

Ref. 4 - p. 12

A. Bornholm disease
B. Guillain-Barré syndrome
C. Louping ill
D. O'nyong-nyong
E. Acute respiratory disease (ARD)

1156. ___ Echovirus types 6, 22
1157. ___ Group A arbovirus
1158. ___ Group B Coxsackie viruses
1159. ___ Adenovirus types 4, 7
1160. ___ Group B arbovirus

Ref. 4 - pp. 527, 586, 494, 876, 641

SECTION IV - VIROLOGY

A. Seasonal incidence of infectious hepatitis
B. Seasonal incidence of rubella
C. Seasonal incidence of poliomyelitis
D. Seasonal incidence of serum hepatitis
E. Seasonal incidence of herpangina

1161.___ Winter and spring
1162.___ Late summer and early fall
1163.___ Autumn and winter
1164.___ Summer
1165.___ Year round   Ref. 4 - pp. 806, 451, 967, 496, 967

A. First stage of smallpox
B. Second stage of smallpox
C. Third stage of smallpox
D. Fourth stage of smallpox
E. Fifth stage of smallpox

1166.___ Macular and papular
1167.___ Pustular
1168.___ Pre-eruptive
1169.___ Crusting
1170.___ Vesicular   Ref. 4 - p. 942

A. Semple type vaccine
B. Sabin vaccine
C. 17D strain vaccine
D. Asian strain vaccine
E. Glycerinated calf lymph vaccine

1171.___ Yellow fever
1172.___ Influenza
1173.___ Poliomyelitis
1174.___ Rabies
1175.___ Smallpox   Ref. 4 - pp. 610, 723-724, 454, 827, 950

A. Mumps
B. Serum hepatitis
C. Influenza
D. Varicella
E. Infectious hepatitis

1176.___ Incubation period 60-160 days
1177.___ Incubation period 14-17 days
1178.___ Incubation period 15-40 days
1179.___ Incubation period 18-21 days
1180.___ Incubation period 1-2 days   Ref. 4 - pp. 967, 919, 967, 759, 704

SECTION IV - VIROLOGY

A. Aedes aegyptii
B. Dermacentor andersoni
C. Phlebotomus papatasii
D. Culex tarsalis
E. Haemagogus spegazzinii

1181. ___ St. Louis encephalitis
1182. ___ Jungle yellow fever
1183. ___ Urban yellow fever
1184. ___ Colorado tick fever
1185. ___ Sandfly fever

Ref. 12 - pp. 1390, 1391, 1391, 1393, 1394

A. P. variolae
B. P. officinale
C. P. bovis
D. P. avium
E. P. muris

1186. ___ Cowpox
1187. ___ Fowlpox
1188. ___ Smallpox
1189. ___ Ectromelia
1190. ___ Vaccinia

Ref. 12 - p. 1254

SELECT THE ITEM FROM COLUMN I WHICH IS UNRELATED TO THE FOUR OTHER ITEMS IN THIS COLUMN, CIRCLE THIS NUMBER. IDENTIFY THE ITEM IN COLUMN II TO WHICH THE FOUR RELATED ITEMS IN COLUMN I ARE ASSOCIATED, CIRCLE THIS LETTER:

I

II

1191.
1. Yellow fever
2. Dengue
3. St. Louis encephalitis
4. Murray Valley encephalitis
5. Western equine encephalomyelitis

A. Examples of arthropod-borne group A virus
B. Examples of arthropod-borne group B virus
C. Examples of arthropod-borne group C virus

Ref. 4 - p. 607

1192.
1. "Black vomit"
2. Faget's sign
3. Hemorrhage
4. High blood pressure
5. Jaundice

A. Symptoms of yellow fever
B. Symptoms of typhus
C. Symptoms of smallpox

Ref. 4 - p. 611

1193.
1. Mumps
2. Influenza
3. Common cold
4. Pharyngoconjunctival fever
5. Bronchiolitis

A. Diseases of the respiratory tract
B. Diseases of the nervous system
C. Diseases of the salivary glands

Ref. 15 - p. 280

1194.
1. RNA viruses
2. Infect a variety of species
3. Produce a characteristic skin lesion
4. Brick-shaped morphology
5. 250 - 330 x 200 - 250 mμ in size

A. Herpesvirus group
B. Poxvirus group
C. Arboviruses

Ref. 4 - p. 16

1195.
1. Variola
2. Alastrim
3. Vaccinia
4. B virus
5. Cowpox

A. Poxviruses
B. Herpesviruses
C. Adenoviruses

Ref. 6 - p. 747

1196.
1. Nucleic acid composition
2. Size
3. Sensitivity to ether
4. Presence of envelope
5. Clinical symptoms of disease

A. Criteria used for classification of viruses
B. Criteria used to determine infectivity of viruses
C. Criteria used in the production of vaccines

Ref. 4 - p. 11

1197.
1. Koplick's spots
2. Rash
3. Cough
4. Stiff neck
5. Fever of 104° to 105°F

A. Symptoms of measles
B. Symptoms of poliomyelitis
C. Symptoms of scarlet fever

Ref. 4 - p. 792

1198.
1. Cervical lymph node enlargement
2. Sore throat
3. Bronchitis
4. High fever
5. Enlarged spleen

A. Symptoms of infectious hepatitis
B. Symptoms of serum hepatitis
C. Symptoms of infectious mononucleosis

Ref. 4 - pp. 996-997

1199.
1. Poliovirus
2. ECHO viruses
3. Mumps virus
4. Coxsackie virus A
5. Coxsackie virus B

A. Picornaviruses
B. Adenoviruses
C. Myxoviruses

Ref. 6 - p. 749

1200.
1. Rash
2. Enlarged parotid glands
3. Increased white cell count in spinal fluid
4. Orchitis
5. Presternal edema

A. Symptoms of measles
B. Symptoms of mumps
C. Symptoms of rubella

Ref. 4 - pp. 759-760

1201.
1. Splenomegaly
2. Vesicles on mucous membranes
3. Fever and malaise
4. Lesions surrounded by erythema
5. Dry cough

A. Symptoms of Q fever
B. Symptoms of varicella
C. Symptoms of phlebotomus fever

Ref. 4 - p. 919

1202.
1. Herpex simplex
2. B virus
3. Varicella-zoster
4. Adeno-associated viruses
5. Burkitt Lymphoma agent

A. Poxviruses
B. Herpesviruses
C. Parvoviruses

Ref. 6 - pp. 747-748

1203.
1. Influenza A
2. Newcastle disease virus
3. Influenza B
4. Influenza C
5. Hemadsorption virus 2

A. Man is the natural host
B. Domestic fowl is the natural host
C. Swine is the natural host

Ref. 4 - p. 686

SECTION IV - VIROLOGY

	I	II
1204.	1. Columbia-SK 2. ECHO 3. EMC 4. FMD 5. Mengo	A. Enteroviruses B. Rhinoviruses C. Adenoviruses Ref. 6 - p. 749
1205.	1. Diffusion 2. Sedimentation 3. Electrophoresis 4. X-ray diffraction 5. Fluorodinitrobenzene method	A. Chemical methods of characterization of virus particles B. Physical methods of characterization of virus particles C. General methods of purification of virus particles Ref. 4 - pp. 27-29
1206.	1. Yellow fever 2. West Nile 3. Dengue 4. Louping ill 5. St. Louis encephalitis	A. Tick serves as vector B. Mosquito serves as vector C. Vector is unknown Ref. 4 - p. 607
1207.	1. Tonsillectomy 2. Pregnancy 3. Alcohol 4. Recent inoculations 5. Trauma	A. Predisposing factors for postinfectious sequelae in measles B. Predisposing factors for paralytic form of poliomyelitis C. Factors responsible for post-vaccinal encephalitis Ref. 4 - pp. 445-446
1208.	1. Yellow fever 2. Dengue fever 3. St. Louis encephalitis 4. West Nile fever 5. O'Nyong-Nyong	A. Primarily encephalitic infections B. Primarily systemic infections C. Human disease not established Ref. 6 - p. 750
1209.	1. La grippe 2. Catarrhal 3. Acute nasopharyngitis 4. Febril catarrh 5. Herpes febrilis	A. Synonyms for herpes simplex B. Synonyms for common cold C. Synonyms for influenza Ref. 4 - p. 689
1210.	1. Largest of the animal viruses 2. Chemically most complex of animal viruses 3. Structurally most complex of animal viruses 4. The presence of RNA 5. They approach the psittacosis group in size and chemical complexity	A. Herpesvirus group B. Myxovirus group C. Poxvirus group Ref. 4 - p. 164

SECTION IV - VIROLOGY

1211.
1. Measles
2. Mumps
3. Rubella
4. Yellow fever
5. Rabies

A. Live attenuated virus vaccine
B. Inactivated virus vaccine
C. Active virus vaccine
Ref. 15 - p. 314

1212.
1. Herpes labialis
2. Cold sores
3. Kaposi's varicelliform eruption
4. Fever blisters
5. Herpes facialis

A. Synonyms for herpes simplex
B. Synonyms for eczema herpeticum
C. Synonyms for varicella
Ref. 4 - p. 898

1213.
1. Produce viremia in certain vertebrates
2. Multiply in certain mosquitoes
3. Transmitted by the infected arthropod
4. Multiply in plants
5. Multiply in certain ticks

A. Psittacosis group
B. Papovaviruses
C. Arboviruses

Ref. 4 - pp. 580-581

1214.
1. Acute respiratory disease of recruits
2. Inculsion conjunctivitis
3. Pharyngoconjunctival fever
4. Conjunctivitis
5. Epidemic keratoconjunctivitis

A. Disease syndromes caused by adenoviruses
B. Disease syndromes caused by Chlamydia
C. Disease syndromes caused by myxoviruses
Ref. 12 - p. 1231

1215.
1. Erythroid leukemia
2. Cowpox
3. Fowlpox
4. Myxoma
5. Molluscum contagiosum

A. Tumor viruses
B. Myxoviruses
C. Poxviruses

Ref. 6 - p. 747

1216.
1. URTI
2. Pleurodynia
3. Pharyngoconjuctival fever
4. Aseptic meningitis
5. Herpangina

A. Adenovirus infections
B. Rhinovirus infections
C. Coxsackie virus infections

Ref. 6 - p. 883

1217.
1. Influenza
2. Parainfluenza
3. Measles
4. Molluscum contagiosum
5. Newcastle disease

A. Poxviruses
B. Myxoviruses
C. Naniviruses

Ref. 6 - pp. 751-752

1218.
1. Poliovirus
2. Equine encephalitis
3. Coxsackie
4. ECHO
5. Mengo

A. Arboviruses
B. Coronaviruses
C. Picornaviruses

Ref. 6 - p. 749

1219.
1. DNA genetic core
2. RNA genetic core
3. Cytopathic effect on CA membrane
4. Guarnieri bodies
5. 200 x 300 nm

A. Characteristics of pox-viruses
B. Characteristics of myxoviruses
D. Characteristics of adenoviruses
Ref. 15 - pp. 412, 416

SECTION IV - VIROLOGY

1220.
1. Poxviruses
2. Adenoviruses
3. Papovaviruses
4. Parvoviruses
5. Picornaviruses

A. Complex
B. Naked double-shelled icosahedral nucleocapsids
C. Naked icosahedral nucleocapsids

Ref. 6 - p. 738

1221.
1. Variola
2. Molluscum contagiosum
3. Ectromelia
4. Herpes simplex
5. Alastrim

A. Host : man
B. Host : rabbit
C. Host : mouse

Ref. 6 - p. 747

1222.
1. Measles virus
2. Mumps virus
3. Influenza A
4. Polioviruses
5. Adenoviruses

A. Enteric viruses
B. Neurotropic viruses
C. Respiratory viruses

Ref. 12 - p. 1214

1223.
1. Influenza A
2. Parainfluenza 1-4
3. Mumps
4. Newcastle disease
5. Measles

A. Myxoviruses
B. Paramyxoviruses
C. Bronchoviruses

Ref. 12 - p. 1312

1224.
1. Eastern equine encephalitis
2. Venezuelan equine encephalitis
3. Western equine encephalitis
4. St. Louis encephalitis
5. Sinbis

A. Major Group A arboviruses
B. Major Group B arboviruses
C. Major Group C arboviruses

Ref. 12 - p. 1377

1225.
1. St. Louis virus
2. Japanese B virus
3. Murray Valley virus
4. Yellow fever virus
5. Ilheus virus

A. Produce central nervous system disease
B. Produce severe systemic disease
C. Produce severe muscle pains and rash

Ref. 12 - p. 1389

ANSWER THE FOLLOWING QUESTIONS BY USING THE KEY OUTLINED BELOW:
A. If 1 and 4 are correct
B. If 2 and 3 are correct
C. If 1, 2 and 3 are correct
D. If all are correct
E. If all are incorrect
F. If some combination other than the above is correct

1226. CYTOPATHIC CHANGES IN A VIRUS-INFECTED CULTURE ARE USUALLY CHARACTERIZED BY THE FOLLOWING ABNORMALITIES:
1. Rounding of cells
2. Formation of syncytial or balloon giant cells
3. Cytoplasmic vacuolation
4. Cytoplasmic granulation

Ref. 14 - p. 83

1227. CLEAR EVIDENCE OF RECOMBINATION HAS BEEN OBTAINED WITH THE FOLLOWING VIRUSES:
1. Poxviruses
2. Influenza
3. Fowl plague
4. Poliomyelitis type 1

Ref. 4 - p. 293

SECTION IV - VIROLOGY

1228. INTERFERON-PRODUCING VIRUSES INCLUDE MEMBERS OF:
1. Arboviruses
2. Poxviruses
3. Myxoviruses
4. Picornaviruses   Ref. 4 - pp. 328-329

1229. THE MOST MARKED PATHOGENIC CHANGES IN YELLOW FEVER ARE OBSERVED IN THE:
1. Liver
2. Spleen
3. Kidney
4. Small intestine   Ref. 4 - p. 611

1230. CHARACTERISTICS OF DENGUE:
1. Severe headache
2. Retrobulbar pain
3. Backache
4. Saddleback fever   Ref. 4 - p. 618

1231. IN STAINED PREPARATIONS OF VIRUS-INFECTED CULTURE, THE FOLLOWING HISTOLOGIC CHANGES CAN BE DETECTED:
1. Nuclear or cytoplasmic inclusions
2. Abnormal accumulations of nucleic acids
3. Dissolution of cell boundaries in the formation of syncytia
4. Nuclear, but not cytoplasmic inclusions
                                Ref. 14 - pp. 84-85

1232. THE SPATIAL ARRANGEMENT OF THE COMPLETE NUCLEIC ACID CHAIN IN A VIRUS MAY BE:
1. Single-stranded
2. Double-stranded
3. Circular in conformation
4. Linear in conformation   Ref. 15 - p. 279

1233. INFECTIOUS HEPATITIS HAS BEEN EXPERIMENTALLY REPRODUCED IN:
1. Guinea pigs
2. Monkeys
3. Chimpanzees
4. Rabbits   Ref. 4 - p. 967

1234. THE SYNONYMS OF ACUTE HERPETIC GINGIVOSTOMATITIS:
1. Acute infectious gingivostomatitis
2. Ulcerative stomatitis
3. Vincent's stomatitis
4. Recurrent stomatitis   Ref. 4 - p. 900

1235. ACUTE RESPIRATORY DISEASE (ARD):
1. Has been described exclusively in military recruits
2. Is usually a relatively mild, grippelike disease
3. The incubation period is from 5 to 6 days
4. Severe throat is a common feature   Ref. 4 - p. 876

1236. LOCALIZED VIRAL DISEASES OF THE SKIN OR MUCOUS MEMBRANES:
1. Herpes simplex
2. Chickenpox
3. Smallpox
4. Herpes zoster   Ref. 15 - p. 280

SECTION IV - VIROLOGY

1237. THE ADENOVIRUS GROUP:
1. Is composed of at least 12 immunologically distinct types of viruses
2. Is composed of at least 45 immunologically distinct types of viruses
3. Lacks pathogenicity for common laboratory animals
4. Types 3, 4 and 7 appear to be the causative agents of epidemic keratoconjunctivitis Ref. 4 - p. 860

1238. RHINOVIRUSES:
1. Have a diameter very similar to that of poliovirus (~30 mµ)
2. Remain infectious after treatment with ethyl ether
3. May be subdivided into H and M rhinoviruses
4. Are DNA viruses Ref. 4 - pp. 547-548

1239. INFLUENZA:
1. Diffuse headache and severe muscular aching of the back are usual in adults
2. Ocular tenderness is uncommon
3. Patient is fatigued and weak
4. Coryza is prominent Ref. 4 - p. 704

1240. GENERALIZED DISEASES IN WHICH VIRUS SPREAD IN THE BODY IS THROUGH THE BLOOD STREAM AND WHICH DO NOT AFFECT SINGLE ORGANS ONLY:
1. Measles
2. Rubella
3. Smallpox
4. Poliomyelitis Ref. 15 - p. 280

1241. CONVALESCENCE FROM INFLUENZA ORDINARILY PROCEEDS RAPIDLY BUT IF THE PATIENT UNDERTAKES FULL ACTIVITY TOO SOON, THE FOLLOWING DISTRESSING FEATURES MAY PRESENT THEMSELVES:
1. Fever
2. Fatigue
3. Mental depression
4. Palpitation Ref. 4 - p. 710

1242. THE DISEASES MOST OFTEN CONFUSED WITH INFECTIOUS MONONUCLEOSIS IN THE YOUNG ADULT INCLUDE:
1. Streptococcal sore throat
2. Exudative tonsillitis of viral etiology
3. Vincent's angina
4. Poliomyelitis Ref. 4 - p. 998

1243. LARGER BACTERIOPHAGES GENERALLY CONSIST OF:
1. A hexagonal "head"
2. A thin protein shell
3. DNA
4. A tail Ref. 6 - p. 828

1244. THE FOLLOWING PROPERTIES ARE CHARACTERISTIC OF THE ADENOVIRUS GROUP:
1. Types 2, 4, 5, 12 and 18 are composed solely of protein and DNA
2. The DNAs of the viruses examined are double-stranded
3. They are relatively stable agents
4. They are inactivated by ether Ref. 4 - pp. 863-864

SECTION IV - VIROLOGY

1245. THE CHEMICAL STRUCTURE OF VACCINIA VIRUS:
1. Protein
2. DNA
3. Phospholipid
4. Neutral fat and carbohydrate   Ref. 4 - p. 954

1246. TOBACCO MOSAIC VIRUS (TMV) IS UNIQUE IN THAT:
1. It was the first virus to be discovered
2. It was the first virus to be purified
3. It was the first virus from which infectious nucleic acid was obtained
4. So far it is the only virus which can be dissociated and reconstituted in vitro   Ref. 4 - p. 67

1247. THE LABORATORY PROCEDURES WHICH HAVE BEEN FOUND TO BE USEFUL IN THE DIAGNOSIS OF SMALLPOX ARE DIRECTED TO THE:
1. Detection of virus
2. Demonstration of specific antigen in focal lesions
3. Demonstration of antibody in the blood during the course of the illness
4. Demonstration of increase in antibody titer during the course of the illness   Ref. 4 - pp. 941-942

1248. THE MEASLES HEMAGGLUTININ:
1. Does not appear to be associated with the viral envelope
2. Is specifically adsorbed by monkey erythrocytes
3. Is not specifically adsorbed by erythrocytes of man
4. There is elution of measles virus from the erthrocytes as with most other myxoviruses   Ref. 4 - p. 789

1249. FOLLOWING INFECTION WITH MUMPS, CERTAIN INDIVIDUALS DEVELOP ANTIBODIES WHICH REACT WITH:
1. Newcastle disease virus
2. Coxsackie viruses
3. Echoviruses
4. Parainfluenza virus   Ref. 4 - p. 762

1250. POLIOVIRUSES ARE INACTIVATED BY:
1. Heat
2. Dessication
3. Formalin
4. Ultraviolet irradiation   Ref. 4 - p. 434

1251. THE FOLLOWING VIRUSES ARE AMONG THE ANIMAL VIRUSES WHICH HAVE YIELDED AN INFECTIOUS RNA:
1. Rhabdoviruses
2. Paramyxoviruses
3. Myxoviruses
4. Picornaviruses   Ref. 15 - p. 295

1252. CONCERNING INFECTIOUS HEPATITIS:
1. The etiologic agent is virus A
2. Virus is in urine during acute phase
3. Onset of symptoms is sudden
4. Infects children and young adults   Ref. 6 - p. 937

1253. INTERFERON IS CHARACTERIZED AS PROTEIN WHICH IS:
1. Acid-stable (pH 2.0)
2. Trypsin-sensitive
3. Nondialyzable
4. Dialyzable   Ref. 15 - p.307

SECTION IV - VIROLOGY

1254. INTERFERON IS PRODUCED BY CELLS IN TISSUE CULTURE WHEN STIMULATED WITH:
1. Viruses
2. Rickettsiae
3. Bacterial endotoxins
4. Synthetic double-stranded polynucleotides
Ref. 15 - p. 307

1255. INCLUSION BODIES WHICH ARE PRODUCED IN THE COURSE OF VIRUS MULTIPLICATION WITHIN CELLS:
1. Are never larger than the individual virus particle
2. Often have an affinity for acid dyes, such as eosin or acid fuchsin
3. May be situated either in the nucleus or in the cytoplasm
4. Never consist of masses of virus particles
Ref. 15 - p. 310

1256. VIRUSES MAY BE TRANSMITTED IN THE FOLLOWING WAYS:
1. Direct contact from person to person
2. By bite of an infected animal host
3. By means of an arthropod vector
4. By means of the alimentary tract   Ref. 15 - p. 311

1257. VIRAL PROTEINS:
1. Make up the principal constituent of all animal virions
2. Are not sole components of capsids
3. Are the minor components of envelopes
4. Vary widely in size from less than 10,000 daltons to more than 150,000

1258. SERUM HEPATITIS:
1. Gamma globulin is not effective
2. Is not contagious
3. Fever is high
4. Affects all ages   Ref. 6 - p. 937

1259. HERPES SIMPLEX VIRUS INFECTIONS:
1. Produce a localized vesicular eruption of the skin and mucous membranes
2. Can be primary or recurrent
3. Primary infection is much more often asymptomatic than not
4. Primary disease occurs in adults   Ref. 6 - p. 858

1260. VIRUSES:
1. Contain a single DNA or RNA
2. Have a protein coat surrounding the nucleic acid
3. Lack constituents fundamental for growth and multiplication
4. Have transfer RNA   Ref. 12 - p. 1014

1261. MORPHOLOGICAL TYPES OF VIRIONS AND OF THEIR COMPONENTS INCLUDE:
1. Naked icosahedral
2. Enveloped icosahedral
3. Naked helical
4. Enveloped helical   Ref. 12 - p. 1021

SECTION IV - VIROLOGY

1262. THE IMPORTANT DIFFERENCES BETWEEN INFECTIONS BY NUCLEIC ACID AND BY VIRIONS ARE:
1. The efficiency of infection with the nucleic acid is much lower
2. The host range is much wider with nucleic acids
3. Infectious nucleic acid can be extracted even from heat-inactivated viruses
4. The infectivity of nucleic acid is unaffected by virus-specific antibodies
Ref. 12 - pp. 1134-1135

1263. INTERFERONS:
1. Consist of small proteins
2. Are unusually stable at low pH
3. Are fairly resistant to heat
4. Are virus-specific
Ref. 12 - p. 1171

1264. VIRAL INACTIVATING AGENTS ARE GROUPED ACCORDING TO THEIR MAIN TARGET AS:
1. Nucleotropic
2. Proteotropic
3. Lipotropic
4. Cytotropic
Ref. 12 - p. 1182

1265. CELLS CAN RESPOND IN THE FOLLOWING WAYS TO VIRAL INFECTION:
1. No apparent change
2. Cytopathic effect and death
3. Hyperplasia followed by death
4. Hyperplasia alone
Ref. 12 - p. 1206

1266. SOME INCLUSION BODIES ARISE AS A CONSEQUENCE OF THE ACCUMULATION OF VIRIONS AND UNASSEMBLED VIRAL SUBUNITS IN:
1. The nucleus
2. The cytoplasm
3. Both the nucleus and the cytoplasm
4. One but not both
Ref. 12 - p. 1207

1267. THE HERPESVIRUSES:
1. Consist of RNA-containing core
2. Consist of an icosahedral capsid
3. Inclusion bodies are intranuclear
4. Site of biosynthesis of subunits is the cytoplasm
Ref. 12 - p. 1238

1268. PATHOGENESIS OF HERPES SIMPLEX VIRUS:
1. The infection persists in a quiescent or latent state in man
2. There is recurrence of activity at irregular intervals
3. The initial infection occurs through a break in the mucous membranes or skin
4. Virus never disseminates into the blood and to distant organs
Ref. 12 - p. 1243

SECTION IV - VIROLOGY

1269. PICORNAVIRUSES INCLUDE:
1. Poliovirus
2. Coxackie virus
3. Echoviruses
4. Rhinoviruses
Ref. 12 - p. 1276

1270. THE GROUP OF VIRUSES NOW DESIGNATED AS ECHOVIRUSES WERE TERMED ENTERIC CYTOPATHIC HUMAN ORPHAN VIRUSES BECAUSE:
1. They were found in the human gastrointestinal tract
2. They produced cytopathic changes in cell cultures
3. They were not clearly associated with disease
4. They were clearly associated with disease
Ref. 12 - p. 1277

1271. THE ADVANTAGES OF THE LIVE ATTENUATED POLIOVIRUS VACCINE:
1. It is easily administered
2. It is relatively inexpensive
3. It produces alimentary tract resistance and therefore confers herd as well as individual immunity
4. Its effectiveness approaches 100 percent
Ref. 12 - p. 1294

1272. THE DISADVANTAGES OF THE LIVE ATTENUATED POLIOVIRUS VACCINE AS PRESENTLY CONSTITUTED ARE:
1. Genetic instability of the type 3 virus employed
2. A possible need for administering each type of virus in individual doses
3. Dissemination of virus to unvaccinated individuals
4. Its effectiveness is only 75 percent
Ref. 12 - p. 1294

1273. NEGRI BODIES:
1. Are characteristic cytoplasmic inclusions of rabies
2. Contain viral particles and virus-specific antigens
3. Contain RNA
4. Contain DNA
Ref. 12 - p. 1335

1274. REOVIRUS:
1. Refers to a group of RNA viruses
2. Infects both the respiratory and the intestinal tracts
3. Produces characteristic cytoplasmic inclusion bodies which contain viral antigens
4. RNA is double-stranded
Ref. 12 - p. 1398

1275. PERTAINING TO MULTIPLICATION OF THE MEASLES VIRUS:
1. It replicates slowly in cell cultures
2. Viral RNA and protein appear to be synthesized solely in the cytoplasm
3. Eosinophilic inclusion bodies develop in both the nuclei and the cytoplasm of giant cells
4. The inclusions are composed of viral particles
Ref. 12 - p. 1358

SECTION IV - VIROLOGY

AFTER EACH OF THE FOLLOWING CASE HISTORIES ARE SEVERAL MULTIPLE CHOICE QUESTIONS ON THE HISTORY. ANSWER BY CHOOSING THE MOST APPROPRIATE ANSWER:

CASE HISTORY - Questions 1276-1280

A 10 year-old boy came home from school because of pain on both sides of the parotid area while eating lunch. The swelling of the glands increased in 2 to 3 days while he remained at home. He complained of a headache and had a low-grade fever. The teacher reported that one of the other students in the classroom had had similar complaints three weeks earlier.

1276. THE DISEASE IS DIAGNOSED AS:
 A. Measles
 B. Common cold
 C. Influenza
 D. Mumps
 E. Infectious mononucleosis
 Ref. 15 - p. 401

1277. THE INCUBATION PERIOD FOR THIS DISEASE IS MOST COMMONLY:
 A. 3 to 5 days
 B. 7 to 12 days
 C. 18 to 21 days
 D. 3 to 5 weeks
 E. 5 to 6 weeks
 Ref. 15 - p. 401

1278. SERIOUS COMPLICATIONS TO BE CONSIDERED IN MUMPS INCLUDE:
 A. Orchitis
 B. Meningoencephalitis
 C. Nephritis
 D. Thyroiditis
 E. All of the above
 Ref. 15 - p. 401

1279. LABORATORY DIAGNOSIS OF MUMPS IS BASED ON:
 A. Recovery of virus from chick embryo inoculation
 B. Recovery of virus from monkey kidney cell culture
 C. Antibody rise in complement fixation test
 D. Antibody rise in hemagglutination inhibition test
 E. All of the above
 Ref. 15 - p. 401

1280. THE MUMPS VIRUS:
 A. Has a DNA core
 B. Agglutinates only chicken red blood cells
 C. Is brick-shaped
 D. Is composed of a helically wound RNA
 E. Possesses strong hemolytic activity
 Ref. 15 - p. 400

SECTION IV - VIROLOGY

CASE HISTORY - Questions 1281-1285

Mr. F. left his place of employment at 3:00 P.M. complaining of headache, fatigue, general achiness and distinct chill. Headache and severe muscular aches ensued and a temperature rise to 103° F. was noted by early morning of the following day. Laryngitis with hoarseness and cough, and substernal soreness were noted. Epistaxis was also noted. Physical examination revealed slight tachycardia and somewhat lower than normal blood pressure. The lymphoid follicles of the soft palate were enlarged and dewy in appearance. The nasal mucous membrane appeared bright red and there were areas of hemorrhage. The patient complained of loss of appetite. Nausea, vomiting or diarrhea was not observed.

1281. THE DISEASE IS DIAGNOSED AS:
    A. Acute respiratory disease
    B. Pharyngitis
    C. Influenza
    D. Common cold
    E. ECHO virus infection
    Ref. 4 - pp. 704-705

1282. THE HIGHEST INCIDENCE IN A PRIMARY EPIDEMIC IS USUALLY IN THE AGE GROUP:
    A. 1-4 years
    B. 5-14 years
    C. Infants
    D. 15-24 years
    E. 25-30 years
    Ref. 4 - p. 712

1283. A SPECIFIC THERAPY FOR THE DISEASE IS:
    A. Sulfonamides
    B. Tetracycline
    C. Penicillin
    D. No specific treatment
    E. Streptomycin
    Ref. 4 - p. 709

1284. THE MOST COMMON COMPLICATION OF THE DISEASE IS:
    A. Pulmonary
    B. Otitis media
    C. Hepatitis
    D. Mycocarditis
    E. Meningoencephalitis
    Ref. 4 - p. 705

1285. THE DURATION OF IMMUNITY TO THIS DISEASE IS:
    A. 4-6 months
    B. Not adequately determined
    C. 1 year
    D. 2 years
    E. 3 years
    Ref. 4 - p. 720

CASE HISTORY - Questions 1286-1290

Mrs. A. has developed a sudden chill, followed by sneezing and coughing; her eyes are red and she is running a temperature. A faint and scattered rash is noticed but Mrs. A. prefers to treat herself before seeking the advice of a physician. Four days after the onset of her illness, the fever and cough have grown steadily worse and the rash appears on her forehead. When seen by her doctor the following day, the exanthem is widespread and maculopapular in appearance. Hemorrhagic forms were noted. The rash disappears on pressure. The patient is also found to be suffering from conjunctivitis with photophobia.

1286. THE DISEASE IS DIAGNOSED AS:
    A. Measles
    B. Exanthem subitum
    C. Hemorrhagic fever
    D. Erythema infectiosum
    E. Vaccinia
    Ref. 4 - p. 792

1287. THE MOST SERIOUS COMPLICATION OF THIS DISEASE IS:
    A. Pneumonia
    B. Encephalomyelitis
    C. Cardiomegaly
    D. Deafness
    E. Epilepsy
    Ref. 4 - p. 793

1288. THE DISEASE IS PARTICULARLY INFECTIVE DURING THE:
A. Catarrhal prodromal stage
B. As long as the rash persists
C. When rash appears
D. 3 days after onset of rash
E. Not infectious at any time
Ref. 4 - p. 795

1289. THE FOLLOWING CHARACTERISTIC FEATURE IS OF ASSISTANCE IN EXCLUDING BACTERIAL INFECTIONS:
A. Leukoplakia
B. Leukopenia
C. Erythroblastemia
D. Erythrocythemia
E. Leukoma
Ref. 4 - p. 793

1290. THE FOLLOWING PERCENTAGE OF SUSCEPTIBLE INDIVIDUALS RESPOND WITH SPECIFIC ANTIBODY TO THE MEASLES VACCINE:
A. 10-20
B. 50
C. 75
D. More than 95
E. Highly variable
Ref. 4 - p. 796

CASE HISTORY - Questions 1291-1295

During the night, Mrs. O. was disturbed by the coughing of her daughter aged 14 months. The next morning the child suffered 2 mild attacks of diarrhea and was very irritable and there was malaise. The child's temperature was 104° F. Aspirins and alcohol sponge baths were administered throughout the day and by evening the temperature had dropped to 102° F. However, the following morning the temperature again rose to 104° F. and by evening was up to 106° F. This rise and fall of temperature continued through the third day. On the morning of the fourth day the child had a fever of 105° F. and then in midafternoon the temperature suddenly turned normal. At this time a fine macular rash was noted on the child's trunk which spread to the neck and arms. This exanthem was thickest on the buttocks and thighs with no rash evident on the face.

1291. THE DISEASE IS DIAGNOSED AS:
A. Measles
B. Rubella
C. Smallpox
D. Roseola subitum
E. Erythema infectiosum
Ref. 4 - p. 811

1292. THE DISEASE IS MOST COMMON DURING:
A. Few months to 2 years
B. 2-4 years
C. 4-6 years
D. 5-7 years
E. All ages
Ref. 4 - p. 811

1293. THIS DISEASE IS:
A. Highly contagious
B. Mildly contagious
C. Not contagious
D. Most prevalent in the winter
E. Most prevalent in the summer
Ref. 4 - p. 811

1294. THE FOLLOWING CHARACTERISTIC FEATURE DISTINGUISHES THIS DISEASE FROM OTHER EXANTHEMATOUS DISORDERS:
A. Sudden onset of fever
B. Fever falls by crisis before the appearance of the rash
C. Enlargement of suboccipital lymph nodes
D. The average incubation period
E. Peak age incidence
Ref. 4 - p. 811

1295. THE SPECIFIC TREATMENT IS:
A. Penicillin
B. Sulfonamides
C. Tetracycline compounds
D. Entirely symptomatic
E. Streptomycin
Ref. 4 - p. 811

SECTION IV - VIROLOGY

CASE HISTORY - Questions 1296-1300

When brought to the hospital, Mr. B. had a temperature of 101° F. His respiration was shallow and his speech interrupted by sighing inspirations. The patient complained of drafts, loud noise and bright lights. Muscle tics and increased activity of muscle reflexes were noted. His pulse was rapid and his pupils dilated. There was increased salivation and excessive perspiration. The patient was extremely nervous, anxious and apprehensive. He had difficulty in swallowing and he was slowly becoming dehydrated because he rejected all fluids as well as solids. Mr. B. had periods of intense excitement at which time he would tear at his bed clothes and bedding. This behavior was suddenly interrupted by a convulsive seizure during which the patient died.

1296. THE DEATH WAS ATTRIBUTED TO:
  A. Aseptic meningitis
  B. Cat-scratch fever
  C. Bulbar poliomyelitis
  D. Rabies
  E. St. Louis encephalitis   Ref. 4 - p. 824

1297. IF THE PATIENT SURVIVES THE EXCITEMENT PHASE OF THIS DISEASE, IT IS FOLLOWED BY:
  A. Paralysis
  B. Hemorrhage
  C. Deafness
  D. Hallucinations
  E. Complete recovery   Ref. 4 - p. 824

1298. AT TIMES THE CLINICAL COURSE MAY BE SIMILAR TO THAT OF:
  A. Phlebotomus fever
  B. Typhoid fever
  C. Poliomyelitis
  D. Diphtheria
  E. Serum hepatitis   Ref. 4 - p. 825

1299. WHEN RABIES IS PRESENT IN A COMMUNITY, ANY DOG OR OTHER PET THAT HAS BITTEN A PERSON MUST BE CONFINED:
  A. For at least 3 days
  B. For at least 5 days
  C. For at least 7 days
  D. For more than 14 days
  E. Confinement is not really necessary
                                           Ref. 4 - p. 834

1300. THE RABIES VACCINATION PROCEDURE IN MAN CONSISTS OF:
  A. Five daily subcutaneous injections in the abdominal wall
  B. Five daily intradermal injections in the abdominal wall
  C. Ten daily subcutaneous injections in the abdominal wall
  D. Fourteen daily subcutaneous injections in the abdominal wall
  E. Fourteen daily intradermal injections in the abdominal wall
                                           Ref. 4 - p. 827

SECTION V - MEDICAL MYCOLOGY

FOR EACH OF THE FOLLOWING MULTIPLE CHOICE QUESTIONS, CHOOSE THE ONE MOST APPROPRIATE ANSWER:

1301. PHYCOMYCOSIS:
- A. Is caused by species of Mucor, Absidia, and Rhizopus
- B. Is rapidly fatal in man
- C. Primarily the brain is invaded
- D. It has been reported most often in patients with uncontrolled diabetes mellitus
- E. All of the above Ref. 6 - pp. 977-978

1302. PATHOGENIC FUNGI ARE MUCH LARGER THAN BACTERIA, EXCEPT FOR:
- A. Actinomyces
- B. Nocardia
- C. Streptomyces
- D. All of the above
- E. None of the above Ref. 6 - p. 956

1303. THE REPRODUCTIVE METHOD WHEREBY FUNGI ASEXUALLY PRODUCE SHORT, OVOID SPORES BY FRAGMENTATION OF HYPHAE IS KNOWN AS:
- A. Chlamydospore formation
- B. Blastospore formation
- C. Conidiospore formation
- D. Sporangiospore formation
- E. Arthrospore formation Ref. 8 - p. 673

1304. SPORANGIOSPORE FORMATION IS OBSERVED IN THE:
- A. Ascomycetes
- B. Phycomycetes
- C. Schizomycetes
- D. Fungi imperfecti
- E. Basidiomycetes Ref. 8 - pp. 668-669

1305. THE DERMATOPHYTES ARE FUNGI WHICH:
- A. Live in the superficial keratinized areas of the body: the skin, hair, and nails
- B. Cause systemic infections
- C. Invariably invade the subcutaneous tissues
- D. Require complex media for growth
- E. Produce morphologically identical macroconidia in all genera Ref. 6 - p. 1000

1306. THE DERMATOPHYTES ARE RESPONSIBLE FOR:
- A. "Ringworm"
- B. "Athelete's foot"
- C. "Jockey itch"
- D. All of the above
- E. None of the above Ref. 6 - p. 1000

1307. MICROSCOPIC IDENTIFICATION OF FUNGI IS BASED ON:
- A. Their staining reaction
- B. The type of spores produced and their arrangement on the hyphae
- C. Their ability to absorb lactophenol-cotton blue
- D. Their solubility in 10 per cent potassium hydroxide
- E. Reaction to eosin-glycerin Ref. 6 - p. 956

1308. PARACOCCIDIOIDES BRASILIENSIS:
- A. Causes a condition known as Gilchrist's disease
- B. Appears in sputum and pus as large, round, thick-walled cells which reproduce by budding
- C. Occurs in tissue, exudates, and sputum as large, round, thick-walled multiple budding cells
- D. Causes three clinical forms: cutaneous, subcutaneous, and systemic
- E. Will never revert to the yeast phase from the filamentous phase Ref. 6 - p. 968

SECTION V - MEDICAL MYCOLOGY

1309. THE CHLAMYDOSPORES WHICH ARE COVERED WITH FINGER-LIKE PROJECTIONS (TUBERCULATE) ARE CHARACTERISTIC AND DIAGNOSTIC FOR:
A. Coccidioides immitis
B. Blastomyces dermatitidis
C. Paracoccidioides brasiliensis
D. Histoplasma capsulatum
E. Geotrichum candidum
Ref. 6 - p. 975

1310. BENIGN, SELF-LIMITING PULMONARY HISTOPLASMOSIS HAS BEEN FOUND TO BE PREVALENT IN INDIVIDUALS RESIDING IN:
A. Central Mississippi Valley
B. Southern California
C. Northeastern United States
D. Northwestern United States
E. None of the above
Ref. 6 - p. 977

1311. THOSE ORGANISMS WHICH DO NOT HAVE A SEXUAL STAGE HAVE BEEN PLACED IN:
A. Phycomycetes
B. Ascomycetes
C. Basidiomycetes
D. Fungi imperfecti
E. None of the above
Ref. 8 - p. 671

1312. CHROMOBLASTOMYCOSIS IS CAUSED BY:
A. Fonsecaea pedrosoi
B. Fonsecaea compactum
C. Cladosporium carrionii
D. Phialophora verrucosa
E. All of the above
Ref. 6 - p. 990

1313. BROWN, SPHERICAL, SEPTATE BODIES FROM PUS ARE DIAGNOSTIC OF:
A. Sporotrichosis
B. Mucormycosis
C. Geotrichosis
D. Maduromycosis
E. Chromoblastomycosis
Ref. 6 - p. 991

1314. THE GENUS CANDIDA REPRODUCES BY:
A. Arthrospore formation
B. Blastospore formation
C. Sexual spores
D. Ascospore formation
E. Sporangiospore formation
Ref. 8 - p. 673

1315. CANDIDA ALBICANS IS NOTEWORTHY AS THE ETIOLOGIC AGENT IN:
A. "Valley fever"
B. "Desert fever"
C. "Desert rheumatism"
D. Thrush
E. San Joaquin fever
Ref. 6 - p. 964

1316. THE SO-CALLED "MADURA FOOT" IS:
A. A slowly progressive unilateral infection of the subcutaneous tissues
B. Characterized by chronicity
C. Characterized by tumefactions
D. Characterized by multiple sinus formation
E. All of the above
Ref. 6 - p. 994

1317. EUROPEAN BLASTOMYCOSIS IS CAUSED BY:
A. Blastomyces dermatitidis
B. Blastomycoides tulanensis
C. Cryptococcus neoformans
D. Coccidioides immitis
E. Paracoccidioides brasiliensis
Ref. 6 - p. 960

1318. CHLAMYDOSPORES FORMED ON THE END OF THE HYPHA ARE CALLED:
A. Intercalary
B. Terminal
C. Lateral
D. Blastospores
E. Arthrospores
Ref. 8 - p. 673

SECTION V - MEDICAL MYCOLOGY

1319. THE ORGANISM ENDEMIC TO THE SAN JOAQUIN VALLEY IN CALIFORNIA:
A. Blastomyces dermatitidis
B. Geotrichum candidum
C. Cryptococcus neoformans
D. Coccidioides immitis
E. Histoplasma capsulatum
Ref. 6 - p. 973

1320. IN TISSUES INFECTED WITH HISTOPLASMA CAPSULATUM:
A. Only the filamentous phase is seen
B. Both the filamentous and yeast phases are seen
C. Thick-walled arthrospores are seen
D. Regional lymph channels are occluded by myceliae
E. The organism is yeast-like   Ref. 6 - p. 974

1321. THE TARGET ORGAN OF CRYPTOCOCCUS NEOFORMANS:
A. Liver
B. Lung
C. Brain
D. Heart
E. Kidney
Ref. 6 - p. 961

1322. THE FOLLOWING FUNGI ARE FOUND IN ENDOTHRIX INFECTIONS OF THE HAIR:
A. Trichophyton schoenleini
B. Trichophyton tonsurans
C. Trichophyton violaceum
D. All of the above
E. None of the above
Ref. 6 - pp. 1007-1008

1323. A FILTRATE OF CULTURE CONTAINING 19 DIFFERENT STRAINS OF THE ORGANISM ELICITS A POSITIVE SKIN REACTION IN 48 HOURS IN SENSITIVE PATIENTS. THE SKIN TEST ANTIGEN IS REFERRED TO AS:
A. Coccidioidin
B. Histoplasmin
C. Paracoccidioidin
D. Blastomycin
E. Tuberculin
Ref. 6 - pp. 969-970

1324. ALL OF THE FOLLOWING ARE DIPHASIC, EXCEPT:
A. Blastomyces dermatitidis
B. Paracoccidioides brasiliensis
C. Cryptococcus neoformans
D. Histoplasma capsulatum
E. Coccidioides immitis
Ref. 6 - p. 960

1325. VESICULAR LESIONS, INDISTINGUISHABLE FROM THE PRIMARY INFECTIONS, WHICH ARISE IN OTHER PARTS OF THE BODY OF AN ALLERGIC INDIVIDUAL INFECTED WITH TRICHOPHYTON ARE REFERRED TO AS:
A. Dermatophytids
B. Trichophytin
C. Eschar
D. Carbuncle
E. Furuncle
Ref. 6 - p. 1010

1326. PRIMARY INFECTION OF THE LUNG WITH ACTINOMYCES ISRAELII RESULTS FROM:
A. Aspiration of the organism from the mouth
B. Bite of an infected mosquito
C. Rubbing in the organism
D. Invasion of the organism through the thoracic area
E. Contaminated fomites   Ref. 8 - p. 5

1327. TISSUE BIOPSIES TAKEN FROM THE FOLLOWING SITES MAY REVEAL THE ORGANISM IN ACTINOMYCOSIS:
A. Sinuses
B. Bone
C. Tongue
D. Lymph nodes
E. All of the above
Ref. 8 - p. 23

SECTION V - MEDICAL MYCOLOGY

1328. ACTINOMYCOSIS PRESENTS A VARIETY OF CLINICAL PICTURES WHICH MUST BE DIFFERENTIATED FROM ALL OF THE FOLLOWING, EXCEPT:
   A. Syphilis
   B. Granuloma inguinale
   C. Typhoid fever
   D. Malaria
   E. Amebiasis
   Ref. 8 - p. 31

1329. SKIN TEST USING THE ORGANISM OR ITS PRODUCTS HAS NOT BEEN SUCCESSFUL IN:
   A. Blastomycosis
   B. Coccidioidomycosis
   C. Actinomycosis
   D. Histoplasmosis
   E. Tuberculosis
   Ref. 8 - p. 31

1330. THE MOST CHARACTERISTIC SINGLE CLINICAL FEATURE COMMON TO NEARLY ALL CASES OF PARACOCCIDIOIDOMYCOSIS IS:
   A. Involvement of lungs
   B. Enlargement of lymph nodes
   C. Enlargement of spleen
   D. Enlargement of liver
   E. Invasion of brain
   Ref. 8 - p. 136

1331. YOUNG IMMATURE FORMS OF COCCIDIOIDES IMMITIS FROM EXUDATES MAY NOT CONTAIN ENDOSPORES AND MAY BE CONFUSED WITH THE NON-BUDDING IMMATURE FORMS OF:
   A. Paracoccidioides brasiliensis
   B. Blastomyces dermatitidis
   C. Geotrichum candidum
   D. Candida albicans
   E. Cryptococcus neoformans
   Ref. 8 - p. 192

1332. PROGRESSIVE COCCIDIOIDOMYCOSIS MUST BE DIFFERENTIATED FROM THE FOLLOWING MYCOSES:
   A. Blastomycosis
   B. Actinomycosis
   C. Cryptococcosis
   D. Sporotrichosis
   E. All of the above
   Ref. 8 - p. 203

1333. THE FOLLOWING PERCENTAGE OF CASES OF PRIMARY HISTOPLASMOSIS IS ASYMPTOMATIC:
   A. 1-5 per cent
   B. 10-25 per cent
   C. 25-50 per cent
   D. 50-75 per cent
   E. At least 95 per cent
   Ref. 8 - p. 224

1334. TORULA HISTOLYTICA IS A SYNONYM FOR:
   A. Blastomyces dermatitidis
   B. Paracoccidioides brasiliensis
   C. Histoplasma capsulatum
   D. Cryptococcus neoformans
   E. Coccidioides immitis
   Ref. 8 - p. 288

1335. THE MOST COMMON LOCALIZED MANIFESTATION OF CUTANEOUS CANDIDIASIS:
   A. Intertrigo
   B. Onychia
   C. Perianal candidiasis
   D. Perléche
   E. "Hairy tongue"
   Ref. 8 - p. 329

1336. THE FOLLOWING SPECIES OF CANDIDA PRODUCES THE CHARACTERISTIC THICK-WALLED, ROUND CHLAMYDOSPORES ON CORN MEAL AGAR:
   A. C. albicans
   B. C. tropicalis
   C. C. krusei
   D. C. parakrusei
   E. C. pseudotropicalis
   Ref. 8 - p. 346

1337. GEOTRICHUM CANDIDUM PRODUCES LESIONS IN THE:
   A. Mouth
   B. Intestinal tract
   C. Lungs
   D. Bronchi
   E. All of the above
   Ref. 8 - p. 365

SECTION V - MEDICAL MYCOLOGY

1338. ASPERGILLOSIS IS CHARACTERIZED BY THE PRESENCE OF GRANULOMATOUS LESIONS IN THE:
   A. Sinuses
   B. Skin
   C. Bronchi
   D. Lungs
   E. All of the above
   Ref. 8 - p. 377

1339. THE MOST COMMON FORM OF SPOROTRICHOSIS:
   A. Skeletal
   B. Mucosal
   C. Cutaneous lymphatic
   D. Disseminated
   E. Visceral
   Ref. 8 - p. 419

1340. PATHOLOGICALLY, MADUROMYCOSIS IS SIMILAR TO:
   A. Candidiasis
   B. Actinomycotic mycetoma
   C. Chromoblastomycosis
   D. Blastomycosis
   E. Sporotrichosis
   Ref. 8 - p. 474

1341. TINEA PEDIS MUST BE DIFFERENTIATED FROM:
   A. Contact dermatitis
   B. Secondary syphilis
   C. Candidiasis
   D. All of the above
   E. None of the above
   Ref. 8 - p. 555

1342. THE FOLLOWING ARE CLASSIFIED AS DERMATOPHYTES, EXCEPT:
   A. Trichophyton tonsurans
   B. Trichophyton violaceum
   C. Candida albicans
   D. Microsporum gypseum
   E. Epidermophyton floccosum
   Ref. 8 - p. 597

1343. SPECIES OF TRICHOPHYTON ATTACK THE:
   A. Hair
   B. Skin
   C. Nails
   D. All of the above
   E. None of the above
   Ref. 8 - p. 598

1344. GEOTRICHOSIS MUST BE DIFFERENTIATED FROM:
   A. Cryptococcosis
   B. Candidiasis
   C. Blastomycosis
   D. All of the above
   E. None of the above
   Ref. 8 - p. 374

1345. ERYTHRASMA IS A CHRONIC MYCOTIC INFECTION OF THE STRATUM CORNEUM CAUSED BY:
   A. Nocardia asteroides
   B. Corynebacterium minutissimum
   C. Aspergillus fumigatus
   D. Actinomyces bovis
   E. Piedraia hortai
   Ref. 8 - p. 652

1346. WHEN CULTURES ARE GROWN ON CYSTINE BLOOD AGAR AT 37° C, THE ORGANISMS ARE FUSIFORM, ROUND, AND OVAL. THEY ARE CHARACTERISTIC OF:
   A. Geotrichum candidum
   B. Sporotrichum schenckii
   C. Hormodendrum pedrosoi
   D. Hormodendrum compactum
   E. Phialophora verrucosa
   Ref. 6 - p. 988

1347. THE FOLLOWING GENUS OF THE FUNGI IMPERFECTI HAS BEEN ISOLATED FROM CASES OF MADUROMYCOSIS:
   A. Madurella
   B. Cephalosporium
   C. Phialophora
   D. Monosporium
   E. All of the above
   Ref. 8 - p. 464

1348. ON SABOURAUD'S AGAR THE FOLLOWING SPECIES DOES NOT HAVE A CREAMY GROWTH:
   A. Candida albicans
   B. Candida krusei
   C. Candida parapsilosis
   D. Candida stellatoidea
   E. Candida guilliermondii
   Ref. 8 - p. 345

SECTION V - MEDICAL MYCOLOGY 147

1349. WHICH IS NOT A SYNONYM FOR TINEA VERSICOLOR ?:
 A. Liver spots
 B. Dermatomycosis furfuracea
 C. Tinea glabrosa
 D. Tinea flava
 E. Pityriasis versicolor tropica   Ref. 8 - p. 644

1350. CRYPTOCOCCUS NEOFORMANS HAS A MARKED PREDILECTION FOR:
 A. Lungs   D. Meninges
 B. Skin   E. None of the above
 C. Liver   Ref. 8 - p. 288

1351. FUNGI RESEMBLE BACTERIA IN:
 A. Their reproductive processes
 B. Their growth characteristics
 C. Their size
 D. Their capacity to cause infectious diseases
 E. The composition and ultrastructure of their cell walls
                         Ref. 12 - p. 976

1352. CRYTOCOCCUS NEOFORMANS:
 A. Sexual forms have been observed
 B. Only certain strains produce capsules
 C. Is not dimorphic
 D. Is dimorphic
 E. Is not encapsulated in tissue   Ref. 12 - p. 982

1353. OF THE SEVERAL MEDICALLY IMPORTANT SPECIES OF CANDIDA, DISEASE IN MAN IS CAUSED MOST FREQUENTLY BY:
 A. C. krusei   D. C. parapsilosis
 B. C. parakrusei   E. C. tropicalis
 C. C. albicans   Ref. 12 - p. 995

1354. CANDIDA PARAPSILOSIS IS THE ORGANISM MOST OFTEN ISOLATED FROM THE FOLLOWING FORM OF CANDIDIASIS:
 A. Thrush   D. Intertriginous candidiasis
 B. Vulvovaginal candidiasis   E. Endocarditis
 C. Bronchopulmonary candidiasis   Ref. 12 - p. 996

1355. ASPERGILLUS FUMIGATUS ACCOUNTS FOR THE FOLLOWING PERCENTAGE OF ASPERGILLOSIS:
 A. 10 per cent   D. 50 per cent
 B. 20 per cent   E. 90 per cent
 C. 40 per cent   Ref. 12 - p. 997

ANSWER THE FOLLOWING QUESTIONS BY USING THE KEY OUTLINED BELOW:
 A. If A is greater in frequency and/or magnitude than B
 B. If B is greater in frequency and/or magnitude than A
 C. If A and B are approximately equal

1356. TINEA CAPITIS CAUSED BY:
 A. Microsporum
 B. Trichophyton   Ref. 8 - p. 573

1357. NUMBER OF BUDS FROM CULTURES GROWN ON BLOOD AGAR AT 37° C:
 A. Paracoccidioides brasiliensis
 B. Blastomyces dermatitidis   Ref. 8 - p. 149

SECTION V - MEDICAL MYCOLOGY

1358. THE SENSITIVITY OF THE FOLLOWING TESTS IN PATIENTS WITH BLASTOMYCOTIC SKIN LESIONS:
   A. Skin test with yeast-phase vaccine
   B. Complement-fixation   Ref. 8 - p. 122

1359. SUSCEPTIBILITY OF COCCIDIOIDOMYCOSIS IN:
   A. Caucasians
   B. American Indians   Ref. 8 - p. 176

1360. PROGRESSIVE DEVELOPMENT OF PRIMARY COCCIDIOIDOMYCOSIS:
   A. Black males
   B. White males   Ref. 8 - pp. 186-187

1361. PROGNOSIS OF:
   A. Primary histoplasmosis
   B. Progressive histoplasmosis   Ref. 8 - p. 264

1362. THE SIGNIFICANCE OF A POSITIVE SKIN TEST TO:
   A. Histoplasmin
   B. Tuberculin   Ref. 8 - p. 262

1363. FREQUENCY OF:
   A. Bronchial candidiasis
   B. Pulmonary candidiasis   Ref. 8 - p. 336

1364. ETIOLOGIC AGENT OF CHROMOBLASTOMYCOSIS:
   A. Fonsecaea pedrosoi
   B. Fonsecaea compactum   Ref. 8 - p. 507

1365. TINEA PEDIS CAUSED BY SPECIES OF:
   A. Trichophyton
   B. Microsporum   Ref. 8 - p. 549

1366. PROGNOSIS OF TINEA CRURIS WHEN DUE TO:
   A. Epidermophyton floccosum
   B. Trichophyton rubrum   Ref. 8 - p. 561

1367. FLUORESCENCE WITH ULTRAVIOLET LIGHT WHEN TINEA BARBAE IS CAUSED BY:
   A. Trichophyton mentagrophytes
   B. Trichophyton rubrum   Ref. 8 - p. 571

1368. PRESENCE OF MULTISEPTATE MACROCONIDIA IN COLONIES OF:
   A. Microsporum canis
   B. Microsporum audouini   Ref. 8 - p. 622

1369. EPIDERMOPHYTON FLOCCOSUM:
   A. Development of microconidia
   B. Development of macroconidia   Ref. 8 - p. 628

1370. ABILITY OF SPECIES OF TRICHOPHYTON TO PRODUCE:
   A. Endothrix type of infection of hair
   B. Ectothrix type of infection of hair   Ref. 8 - p. 598

1371. THE CURE OF TINEA PEDIS CAUSED BY:
   A. Trichophyton mentagrophytes
   B. Trichophyton rubrum   Ref. 8 - p. 555

1372. PATHOGENICITY OF CANDIDA ALBICANS:
   A. Superficial and epithelial
   B. Deeper involvement   Ref. 8 - p. 349

SECTION V - MEDICAL MYCOLOGY

1373. MODE OF TRANSMISSION OF CRYPTOCOCCOSIS:
A. From infected human
B. From infected animal
Ref. 8 - p. 290

1374. SITE OF FUNGI IN CHROMOBLASTOMYCOSIS:
A. Dermis
B. Epidermis
Ref, 8 - p. 522

1375. GEOGRAPHICAL DISTRIBUTION OF HISTOPLASMOSIS:
A. St. Lawrence River Valley
B. Central Mississippi Valley
Ref. 8 - p. 218

ANSWER THE FOLLOWING QUESTIONS BY USING THE KEY OUTLINED BELOW:
A. If both statement and reason are true and related cause and effect
B. If both statement and reason are true but not related cause and effect
C. If the statement is true but the reason is false
D. If the statement is false but the reason is true
E. If both statement and reason are false.

1376. Cervicofacial actinomycosis has the best prognosis BECAUSE it is the most common form of the disease. Ref. 8 - p. 3

1377. The physical signs in the early stages of pulmonary actinomycosis resemble those of tuberculosis BECAUSE in both diseases the primary sites of infection are found most frequently at the lung bases.
Ref. 8 - p. 5

1378. The edge of a lesion of cutaneous blastomycosis shows extraordinary hypertrophy and hyperplasia of the epidermis BECAUSE of the chronic inflammatory reaction incited by the fungus.
Ref. 8 - p. 110

1379. Paracoccidioides brasiliensis was placed in the genus Blastomyces BECAUSE of the similarities to B. dermatitidis in tissues and in culture.
Ref. 8 - p. 144

1380. Extreme caution should be exercised when transferring or examining the growth of Coccidioides immitis BECAUSE the arthrospores are highly infectious. Ref. 8 - p. 195

1381. Primary histoplasmosis resembles primary coccidioidomycosis BECAUSE both are highly malignant infections of the lungs.
Ref. 8 - p. 224

1382. Histoplasma capsulatum can be easily demonstrated in fresh and unstained preparations of clinical materials BECAUSE the fungus appears as small oval bodies in large mononuclear cells.
Ref. 8 - p. 242

1383. The examination of spinal fluid is more important than histologic examination in the case of cryptococcosis BECAUSE it is found most often in the central nervous system or meninges.
Ref. 8 - p. 305

1384. C. albicans vulvovaginitis is common particularly in diabetes BECAUSE the infection is apparently related to the large amounts of sugar present in the blood and urine. Ref. 8 - p. 328

SECTION V - MEDICAL MYCOLOGY

1385. Demonstration of Geotrichum by direct examination is essential for diagnosis BECAUSE Geotrichum not infrequently is found in the sputum in association with Friedlander's bacillus.
Ref. 8 - p. 373

1386. Aspergillosis has been recognized as an occupational disease BECAUSE species of Aspergillus are ubiquitous. Ref. 6 - p. 1014

1387. Only Candida albicans is considered to be pathogenic BECAUSE other species are not encountered in pathologic conditions.
Ref. 6 - p. 962

1388. Cryptococcosis is a true "blastomycosis" BECAUSE the organism appears in tissue as a budding fungus. Ref. 8 - p. 288

1389. The term "European blastomycosis" used as a synonym for cryptococcosis is inappropriate BECAUSE the infection is not limited to Europe but occurs in many parts of the world.
Ref. 8 - p. 288

1390. A classification for dermatophytoses on an etiologic basis is satisfactory BECAUSE a single species of a dermatophyte can cause only a single clinical manifestation. Ref. 8 - p. 548

1391. The name "chromoblastomycosis" is appropriate for the disease BECAUSE the fungi form buds in tissue and the lesions have a characteristic color. Ref. 8 - p. 503

1392. Dermatophytid reactions in man result from spread of fungus elements through the blood stream BECAUSE the "id" lesions disappear after clearing of the fungi from the primary lesions.
Ref. 8 - p. 589

1393. Geotrichosis, like candidiasis, is believed to be endogenous in origin BECAUSE the fungus frequently can be found in the mouths and intestinal tracts of normal individuals. Ref. 8 - p. 365

1394. The candidids or "levurids" do not have the same clinical characteristics as the dermatophytids BECAUSE these are not sterile lesions.
Ref. 8 - p. 329

1395. The disease due to P. brasiliensis is often referred to as paracoccidioidal granuloma BECAUSE the tissue form of the fungus causing the infection was thought at first to resemble the organism that causes coccidioidomycosis. Ref. 8 - p. 134

ANSWER THE FOLLOWING QUESTIONS BY USING THE KEY OUTLINED BELOW:
A. If 1 and 4 are correct
B. If 2 and 3 are correct
C. If 1, 2 and 3 are correct
D. If all are correct
E. If all are incorrect
F. If some combination other than the above is correct

1396. BLASTOMYCES DERMATITIDIS:
1. Is dimorphic
2. Grows only in enriched media
3. Appears as encapsulated yeast cells in tissue
4. Appears as thick-walled, multinucleated spherical cells in tissue
Ref. 12 - p. 985

SECTION V - MEDICAL MYCOLOGY

1397. IN THE SKIN TEST WITH HISTOPLASMIN, CROSS REACTIONS OCCUR WITH:
    1. Tuberculin
    2. Coccidioidin
    3. Blastomycin
    4. Oidiomycin   Ref. 12 - p. 988

1398. THE ORGANISM THAT CAUSES SOUTH AMERICAN BLASTOMYCOSIS:
    1. Appears in infected tissue as large, spherical or oval yeast cells
    2. Measures 10 to 30µ in diameter in tissue
    3. Multiple buds sprout from a single mother cell in tissue
    4. Grows as a mycelium with chlamydospores at room temperature
                          Ref. 12 - **pp.** 993-994

1399. SUBCUTANEOUS MYCOSES INCLUDE:
    1. Sporotrichosis
    2. Chromoblastomycosis
    3. Maduromycosis
    4. Candidiasis   Ref. 12 - p. 1000

1400. THE DERMATOPHYTES WHICH ATTACK ONLY HAIR AND SKIN:
    1. Microsporum
    2. Epidermophyton
    3. Trichophyton
    4. Keratinomyces   Ref. 12 - p. 1002

1401. THE FOLLOWING ARE SYNONYMS FOR TINEA PEDIS:
    1. Jockey itch
    2. Gym itch
    3. Favus honeycomb ringworm
    4. Scaly ringworm   Ref. 8 - p. 549

1402. TINEA CAPITIS:
    1. Is an infection of the scalp and hair
    2. Is caused by species of Trichophyton and Microsporum
    3. Has a limited geographic distribution
    4. Is caused most commonly in children by Microsporum
                          Ref. 8 - p. 573

1403. THE ASEXUAL SPORES (THALLOSPORES) INCLUDE:
    1. Blastospores
    2. Chlamydospores
    3. Arthrospores
    4. Ascospores   Ref. 8 - p. 668

1404. THE TRUE FUNGI ARE DIVIDED INTO:
    1. Phycomycetes
    2. Ascomycetes
    3. Basidiomycetes
    4. Fungi imperfecti   Ref. 8 - p. 665

1405. GROSSLY, THE MYCELIAL CULTURES OF THE FOLLOWING FUNGI ARE IDENTICAL:
    1. Coccidioides immitis
    2. Histoplasma capsulatum
    3. Blastomyces dermatitidis
    4. Cryptococcus neoformans   Ref. 8 - p. 250

SECTION V - MEDICAL MYCOLOGY

SELECT THE ITEM FROM COLUMN I WHICH IS UNRELATED TO THE FOUR OTHER ITEMS IN THIS COLUMN, CIRCLE THIS NUMBER. IDENTIFY THE ITEM IN COLUMN II TO WHICH THE FOUR RELATED ITEMS IN COLUMN I ARE ASSOCIATED, CIRCLE THIS LETTER:

	I	II
1406.	1. Valley fever 2. Desert rheumatism 3. San Joaquin fever 4. Coccidioidal granuloma 5. Darling's disease	A. Histoplasmosis B. Coccidioidomycosis C. Cryptococcosis Ref. 8 - p. 171
1407.	1. Aerobic 2. Acid-fast 3. Partially acid-fast 4. Isolated from soil 5. Gram-positive	A. Mycobacterium B. Nocardia C. Actinomyces Ref. 8 - p. 38
1408.	1. Typical chlamydospores 2. Fermentation of glucose and maltose to acid and gas 3. Pseudomycelium 4. Development of blastospores 5. Fragmentation of hyphae	A. Candida B. Geotrichum C. Sporotrichum Ref. 6 - p. 963
1409.	1. Fluoresces under Wood's light 2. Club-shaped macroconidia in clusters 3. Tinea capitis 4. Fusiform macroconidia 5. Zoophilic	A. Trichophyton B. Epidermophyton C. Microsporum Ref. 6 - pp. 1002-1003
1410.	1. Reddish pigment formation 2. Long, slender and pencil-shaped macroconidia 3. Elongated, clavate, and single-celled microconidia 4. Causes recalcitrant lesions of skin and nails 5. Ectothrix infection of the hair	A. Trichophyton rubrum B. Trichophyton mentagrophytes C. Trichophyton violaceum Ref. 6 - pp. 1006-1007
1411.	1. Tinea unguium 2. Tinea glabrosa 3. Tinea pedis 4. Tinea barbae 5. Tinea capitis	A. Infection of skin B. Infection of hair C. Infection of nail Ref. 6 - pp. 1010-1011
1412.	1. Dermatophytosis 2. Epidermatophytosis 3. Dermatomycosis 4. Eczema marginatum 5. "Athlete's foot"	A. Tinea pedis B. Tinea unguium C. Tinea cruris Ref. 8 - p. 549
1413.	1. Chains of dark green spherical conidia 2. Hyphal mass in lung 3. Spore-bearing hyphae like a brush 4. Ubiquitous in nature 5. Dark green, powdery appearance	A. Mucor B. Penicillium C. Aspergillus Ref. 6 - pp. 1014-1015

SECTION V - MEDICAL MYCOLOGY

1414.
1. Blastospores
2. Chlamydospores
3. Arthrospores
4. Zygospore
5. Conidia

A. Sexual spores
B. Asexual spores
C. Hyphae

Ref. 8 - pp. 668-669

1415.
1. Thrush
2. Perlèche
3. Vulvovaginitis
4. Vaginitis
5. Erythrasma

A. Candida albicans
B. Coccidioides immitis
C. Histoplasma capsulatum

Ref. 6 - p. 964

MATCH EACH OF THE FIVE ITEMS LISTED BELOW WITH THE MOST APPROPRIATE STATEMENTS OR CHOICES. EACH ITEM MAY BE USED ONLY ONCE:

A. Thrush
B. Darling's disease
C. Paracoccidioidal granuloma
D. Phycomycosis
E. Streptothricosis

1416. ___ Mucor sp.
1417. ___ P. brasiliensis
1418. ___ Candida albicans
1419. ___ Histoplasma capsulatum
1420. ___ Nocardia asteroides

Ref. 8 - pp. 409, 134, 325, 218, 38

A. Verrucous dermatitis
B. "Levurids"
C. Torulosis
D. Gram-positive, cigar-shaped cells
E. Mandura foot

1421. ___ Sporotrichum schenckii
1422. ___ Allescheria boydii
1423. ___ Fonsecaea pedrosoi
1424. ___ Candida albicans
1425. ___ Cryptococcus neoformans

Ref. 8 - pp. 441, 458, 503, 507, 329, 288

A. Tinea cruris
B. Tinea favosa
C. Tinea capitis
D. Tinea imbricata
E. Tinea corporis

1426. ___ Tinea glabrosa
1427. ___ Concentric rings on body
1428. ___ Gym itch
1429. ___ "Black dot"
1430. ___ Scutula

Ref. 8 - pp. 563, 568, 560, 575, 579

SECTION V - MEDICAL MYCOLOGY

    A. Lepothrix
    B. "Oidiomycin" vaccine
    C. Piedra
    D. "Dematium" type
    E. Liver spots

1431. ___ Tinea nodosa
1432. ___ Tinea versicolor
1433. ___ Trichophytin
1434. ___ Trichomycosis axillaris
1435. ___ Tinea nigra palmaris

Ref. 8 - pp. 632, 644, 593, 639, 496

    A. Black piedra
    B. Kerion
    C. Favus
    D. White piedra
    E. Tinea nigra

1436. ___ Microsporum canis
1437. ___ Trichosporon cutaneum
1438. ___ Cladosporium werneckii
1439. ___ Trichophyton schoenleinii
1440. ___ Piedraia hortai

Ref. 12 - pp. 1006, 1008, 1008, 1008, 1008

AFTER EACH OF THE FOLLOWING CASE HISTORIES ARE SEVERAL MULTIPLE CHOICE QUESTIONS BASED ON THE HISTORY. ANSWER BY CHOOSING THE MOST APPROPRIATE ANSWER:

CASE HISTORY - Questions 1441-1445

A white male, 20 years of age, who had moved from Kentucky to New York, came into the hospital complaining of a mild cough, fever, loss of weight and strength. There were rales in the lungs. The symptoms were apparent only recently. The X-ray showed multiple lesions scattered throughout both lung fields. The hilar lymph nodes were enlarged Laboratory findings showed hypochromic anemia and leukopenia.

1441. THE HISTORY AND SYMPTOMS SUGGEST THE INFECTION TO BE MOST LIKELY:
    A. Histoplasmosis
    B. Coccidioidomycosis
    C. Cryptococcosis
    D. Pulmonary candidiasis
    E. Geotrichosis
    Ref. 8 - pp. 218-227

1442. IF THERE IS HEMOPTYSIS, IT SHOULD SUGGEST THE POSSIBILITY OF A COEXISTING INFECTION OF:
    A. Tuberculosis
    B. Blastomycosis
    C. Coccidioidomycosis
    D. Cryptococcosis
    E. Bacterial pneumonia
    Ref. 8 - p. 234

1443. TO ESTABLISH THE DIAGNOSIS, THE ORGANISM SHOULD BE DEMONSTRATED IN OR CULTIVATED FROM THE:
    A. Blood
    B. Sternal bone marrow
    C. Sputum
    D. Lymph nodes
    E. Any one of the above
    Ref. 8 - p. 263

1444. A POSITIVE SKIN TEST WITH THE STANDARDIZED HISTOPLASMIN INDICATES:
    A. Immunity
    B. Susceptibility
    C. A past or present infection
    D. Chronicity of the infection
    E. None of the above
    Ref. 8 - p. 262

SECTION V - MEDICAL MYCOLOGY

1445. THE DRUG OF CHOICE FOR HISTOPLASMOSIS:
- A. Iodides
- B. Heavy metals
- C. Sulfonamides
- D. Penicillin
- E. Amphotericin B

Ref. 8 - p. 265

CASE HISTORY - Questions 1446-1450

A young soldier, 25 years of age, was examined because he came into the hospital with a high fever and had experienced night sweats. He had lost some weight and the dyspnea had increased. His sputum was somewhat purulent and contained bits of blood. There was dullness and the breath sounded altered. There was a small discharging sinus over the thorax. X-ray films showed dense masses with irregular outlines projecting from the hilum. The lesion was unilateral.

1446. IF THE CONDITION IS BLASTOMYCOSIS, PRESUMPTIVE DIAGNOSIS CAN BE MADE BY:
- A. A complete blood chemistry
- B. Skin test with Blastomyces vaccine
- C. X-ray of the lesion
- D. Gram stain of the smear from the draining lesion
- E. None of the above

Ref. 8 - p. 102

1447. THE SOURCE OF BLASTOMYCOSIS:
- A. It spreads from man to man
- B. It spreads from animal to man
- C. B. dermatitidis is ubiquitous in nature
- D. Man presumably derives the infection from some exogenous source
- E. Man derives it from endogenous source

Ref. 8 - p. 85

1448. IF THE ORGANISM OBTAINED FROM THE LESION AND PUS IS EXAMINED IN 10 PER CENT POTASSIUM HYDROXIDE UNDER THE MICROSCOPE, B. DERMATITIDIS APPEARS AS:
- A. Single or budding spherical cells with a thick refractile wall
- B. Single budding spherules surrounded by a wide capsule
- C. Multiple budding cells
- D. Round, thick-walled endospore-filled spherules
- E. Small oval bodies within large mononuclear cells

Ref. 8 - p. 103

1449. IF THE MATERIAL FROM THE LESION IS CULTURED ON BLOOD AGAR AT 37° C:
- A. B. dermatitidis grows slowly
- B. The colonies are wrinkled and waxy
- C. Microscopic examination shows organisms to be identical to those from infected lesions
- D. All of the above
- E. None of the above

Ref. 8 - p. 107

1450. SYSTEMIC BLASTOMYCOSIS MUST BE DIFFERENTIATED FROM:
- A. Tuberculosis
- B. Syphilis
- C. Actinomycosis
- D. Coccidioidomycosis
- E. All of the above

Ref. 8 - p. 123

SECTION VI - MEDICAL PARASITOLOGY

FOR EACH OF THE FOLLOWING MULTIPLE CHOICE QUESTIONS, CHOOSE THE ONE MOST APPROPRIATE ANSWER:

1451. SYMBIOSIS IS MOST ACCURATELY DESCRIBED AS A:
A. Relationship in which both organisms benefit
B. Permanent association, in as much as both organisms cannot exist independently
C. Relationship in which only the host benefits
D. Relationship in which one is benefited and the other is unaffected
E. Relationship in which one species depends upon another for survival
Ref. 1 - p. 1

1452. FACULTATIVE PARASITES:
A. Can lead both a free and parasitic life
B. Are completely dependent upon the host
C. Establish themselves in hosts in which they do not ordinarily live
D. Are free-living during part of their existence and seek their hosts intermittently
E. Remain on or in the body of the host for their entire lives
Ref. 1 - p. 1

1453. THE FOLLOWING CLASSES OF PROTOZOA CONTAIN IMPORTANT PARASITES OF MAN:
A. Rhizopoda
B. Ciliata
C. Mastigophora
D. Sporozoa
E. All of the above
Ref. 11 - pp. 13-14

1454. A SPURIOUS PARASITE THAT HAS PASSED THROUGH THE ALIMENTARY TRACT INFECTING MAN IS KNOWN AS A:
A. Coprozoic parasite
B. Pseudoparasite
C. Incidental parasite
D. Temporary parasite
E. None of the above
Ref. 1 - p. 1

1455. THE ADULT OR SEXUALLY MATURE STAGE OF THE PARASITE OCCURS IN THE:
A. First intermediate host
B. Final or definitive host
C. Second intermediate host
D. Insect vector
E. None of the above
Ref. 1 - p. 2

1456. THE CHIEF SOURCE OF MOST PARASITIC DISEASES OF MAN:
A. Mosquitoes
B. Birds
C. Dogs
D. Cattle
E. Man
Ref. 1 - p. 3

1457. THE FOLLOWING PROTOZOAN OCCASIONALLY UNDERGOES CONJUGATION:
A. Trichomonas tenax
B. Entamoeba histolytica
C. Iodamoeba bütschlii
D. Dientamoeba fragilis
E. Balantidium coli
Ref. 9 - p. 22

1458. THE FOLLOWING ARTHROPOD, WHICH IS MEDIUM-SIZED, YELLOWISH TO MEDIUM-BROWN OR DARK-BROWN IN COLOR, WITH A DISTINCT CONSTRICTION BETWEEN CEPHALOTHORAX AND ABDOMEN, IS RESPONSIBLE FOR A LOCAL NECROTIZING ULCER:
A. Centruroides
B. Latrodectus
C. Loxosceles
D. Sarcoptes
E. Scolopendra
Ref. 9 - p. 325

SECTION VI - MEDICAL PARASITOLOGY

1459. PATHOGENIC PROTOZOA AFFECT THE HOST BY THEIR:
    A. Multiplication
    B. Invasion
    C. Destruction of cells
    D. Toxic or enzymatic activities
    E. All of the above
    Ref. 1 - p. 15

1460. ALL OF THE FOLLOWING AMEBAE LIVE IN THE LARGE INTESTINE, EXCEPT:
    A. Entamoeba histolytica
    B. Entamoeba coli
    C. Entamoeba gingivalis
    D. Endolimax nana
    E. Iodamoeba bütschlii
    Ref. 1 - p. 17

1461. THE MOST IMPORTANT PATHOGENIC PARASITIC AMEBA OF MAN:
    A. Endolimax nana
    B. Iodamoeba bütschlii
    C. Entamoeba histolytica
    D. Dietamoeba fragilis
    E. Entamoeba coli
    Ref. 1 - p. 17

1462. THE TROPHOZOITE OF ENTAMOEBA HISTOLYTICA:
    A. Is usually larger than that of E. coli
    B. Nuclear membrane is lined with coarse granules of chromatin
    C. Red blood cells are always present
    D. May have more than one nucleus
    E. Ectoplasmic pseudopodia are extended rapidly
    Ref. 1 - p. 21

1463. THE USUAL HABITAT OF THE TROPHOZOITE OF E. HISTOLYTICA IS THE:
    A. Liver
    B. Small intestine
    C. Stomach
    D. Colon
    E. Gallbladder
    Ref. 1 - p. 22

1464. THE NATURAL HOST OF GIARDIA LAMBLIA:
    A. Rat
    B. Man
    C. Cow
    D. Dog
    E. Bird
    Ref. 1 - p. 40

1465. TRICHOMONAS VAGINALIS:
    A. The trophozoite is one of the most resistant of the parasitic protozoa
    B. Clinicians generally regard T. vaginalis as the causative agent of a persistent vaginitis
    C. The basis of treatment is the restoration of normal conditions in the vagina
    D. All of the above
    E. None of the above
    Ref. 1 - pp. 41-42

1466. IN TRICHOMONAS VAGINITIS THE FACTORS THAT DETERMINE INFECTION INCLUDE:
    A. The size of the dose
    B. The bacterial flora
    C. The vaginal cells
    D. Physiologic status of the vagina
    E. All of the above
    Ref. 1 - p. 42

1467. THE FOLLOWING PATHOGENIC SPECIES OF TRYPANOSOMA IS PRESENT IN AMERICA:
    A. T. gambiense
    B. T. rhodesiense
    C. T. cruzi
    D. All of the above
    E. None of the above
    Ref. 1 - p. 45

SECTION VI - MEDICAL PARASITOLOGY

1468. THE PRINCIPAL INSECT VECTORS OF TRYPANOSOMA RHODESIENSE:
 A. Reduviid bugs
 B. Sandflies
 C. Tsetse flies
 D. Cockroaches
 E. Mosquitoes
 Ref. 1 - p. 54

1469. CHAGAS' DISEASE (AMERICAN TRYPANOSOMIASIS) IS CAUSED BY:
 A. T. rhodesiense
 B. T. gambiense
 C. T. cruzi
 D. T. rangeli
 E. None of the above
 Ref. 1 - p. 55

1470. KALA-AZAR IS CAUSED BY:
 A. T. rhodesiense
 B. T. gambiense
 C. T. cruzi
 D. T. brucei
 E. None of the above
 Ref. 1 - p. 61

1471. THE PRIMARY PATHOLOGIC CHANGE IN MALARIA:
 A. Reactions to toxins
 B. Destruction of erythrocytes
 C. Destruction of lymphocytes
 D. Anoxemic impairment of tissues
 E. Venous congestion
 Ref. 1 - p. 84

1472. MALARIAL RELAPSES OCCURRING MOSTLY WITHIN FIRST TWO YEARS AND UNUSUALLY AFTER THREE YEARS ARE DUE TO:
 A. P. falciparum
 B. P. vivax
 C. P. malariae
 D. P. berghei
 E. P. cynomolgi
 Ref. 9 - p. 105

1473. THE MAIN SYMPTOMS OF ACUTE VIVAX OR QUARTAN MALARIA:
 A. Chills and fever
 B. Nausea and vomiting
 C. Malaise
 D. Muscular pains
 E. All of the above
 Ref. 9 - p. 105

1474. "BLACKWATER FEVER" RESULTS FROM:
 A. Splenic enlargement
 B. Liver enlargement
 C. Anaphylaxis
 D. Intravascular hemolysis
 E. Melanosarcoma
 Ref. 1 - p. 90

1475. THE IMPORTANT HUMAN HELMINTHS INCLUDE:
 A. Nematodes
 B. Cestodes
 C. Trematodes
 D. Only A and B
 E. All of the above
 Ref. 11 - p. 93

1476. TRICHURIS TRICHIURA MAINTAINS ITS POSITION IN THE INTESTINAL TRACT BY:
 A. Anchorage with its attenuated anterior portion
 B. Penetration into tissues
 C. Temporary attachment or retention in the folds of the mucosa
 D. Attachment with its cutting plates
 E. None of the above
 Ref. 1 - p. 115

1477. THE FOLLOWING INTESTINAL NEMATODES ARE INFECTIVE FOR MAN IN THE EGG STAGE, EXCEPT:
 A. Enterobius vermicularis
 B. Trichuris trichiura
 C. Ascaris lumbricoides
 D. Trichinella spiralis
 E. None of the above
 Ref. 11 - p. 98

1478. OXYURIASIS IS CAUSED BY:
 A. Ascaris lumbricoides
 B. Dracunculus medinensis
 C. Enterobius vermicularis
 D. Trichuris trichiura
 E. Necator americanus
 Ref. 1 - p. 129

SECTION VI - MEDICAL PARASITOLOGY 159

1479. ASCARIS PNEUMONITIS IS CHARACTERIZED BY:
A. Dyspnea
B. Cough
C. Eosinophila
D. All of the above
E. None of the above
Ref. 11 - p. 107

1480. THE USUAL PORTAL OF ENTRY FOR STRONGYLOIDES STERCORALIS IS THE:
A. Intestine
B. Skin
C. Nose
D. Mouth
E. Umbilicus
Ref. 11 - p. 123

1481. LARVAE OF TRICHINELLA SPIRALIS ARE KILLED AT TEMPERATURES OF ABOUT:
A. 20° C
B. 30° C
C. 40° C
D. 45° C
E. 55° C
Ref. 11 - p. 327

1482. IN UNCOMPLICATED HEAVY TRICHURIS INFECTIONS, THE MOST FREQUENT COMPLAINT IS:
A. Abdominal or epigastric pain
B. Muscular ache
C. Cough
D. Neuralgia
E. Extreme weakness
Ref. 1 - p. 115

1483. THE SEQUENCE OF AN ATYPICAL PNEUMONITIS, FOLLOWED IN A FEW WEEKS BY WATERY DIARRHEA, EPIGASTRIC DISTRESS AND EOSINOPHILIA IS SUGGESTIVE OF:
A. Amebic dysentery
B. Strongyloidiasis
C. Trichinosis
D. Trichuriasis
E. Enterobiasis
Ref. 1 - p. 120

1484. CUTANEOUS LARVA MIGRANS RESULTING FROM EXPOSURE TO THE FOLLOWING PARASITE HAS A WIDESPREAD DISTRIBUTION THROUGH-OUT THE EASTERN COASTAL AREAS OF THE UNITED STATES:
A. Ancylostoma caninum
B. Uncinaria stenocephala
C. Necator americanus
D. Ancylostoma braziliense
E. Ancylostoma duodenale
Ref. 9 - p. 264

1485. THE MOST COMMON HELMINTHIC PARASITE OF MAN IN THE U.S.A.:
A. Enterobius vermicularis
B. Trichuris trichiura
C. Ascaris lumbricoides
D. Trichinella spiralis
E. Ancylostoma duodenale
Ref. 1 - p. 131

1486. THE USUAL HABITAT OF THE PINWORM IS THE:
A. Stomach
B. Duodenum
C. Appendix
D. Cecum
E. Anus
Ref. 1 - p. 129

1487. ASCARIASIS IS MOST PREVALENT IN THE FOLLOWING AGE GROUP:
A. Under 1 year
B. 1 to 4 years
C. 5 to 9 years
D. 10 to 15 years
E. 20 to 30 years
Ref. 1 - p. 135

1488. THE SLENDER RHABDITIFORM LARVAE OF THE FOLLOWING HELMINTH MOVE ABOUT IN WATER AND ARE INGESTED BY SPECIES OF CYCLOPS:
A. Diphyllobothrium latum
B. Dracunculus medinensis
C. Wuchereria bancrofti
D. Schistosoma mansoni
E. Clonorchis sinensis
Ref. 1 - p. 157

SECTION VI - MEDICAL PARASITIOLGY

1489. NOCTURNAL PERIODICITY IS MOST CHARACTERISTIC OF THE MICROFILARIAE OF THE FOLLOWING SPECIES IN THE WESTERN HEMISPHERE:
   A. Onchocerca volvulus
   B. Loa Loa
   C. Acanthocheilonema perstans
   D. Wuchereria bancrofti
   E. Mansonella ozzardi
   Ref. 1 - p. 142

1490. THE ADULT WORMS OF WUCHERERIA BANCROFTI ARE LOCATED IN THE:
   A. Blood
   B. Lymphatics
   C. Intestinal tract
   D. Heart
   E. Spleen
   Ref. 1 - p. 144

1491. THE PRINCIPAL VECTOR IN THE WESTERN HEMISPHERE FOR WUCHERERIASIS:
   A. Aedes
   B. Culex
   C. Anopheles
   D. Mansonia
   E. None of the above
   Ref. 1 - p. 146

1492. THE ADULT WORMS OF ONCHOCERCA VOLVULUS ARE USUALLY FOUND IN THE:
   A. Liver
   B. Blood
   C. Lymph
   D. Subcutaneous tissues
   E. Lymphatics
   Ref. 1 - p. 150

1493. OBSERVATION OF WORMS UNDER THE CONJUNCTIVA, CALABAR SWELLINGS AND EOSINOPHILIA ARE DIAGNOSTIC FOR:
   A. Acanthocheilonemiasis
   B. Onchocercosis
   C. Loasis
   D. Bancroft's filariasis
   E. Ascariasis
   Ref. 1 - p. 155

1494. THE FOLLOWING COMMON TAPEWORMS OF MAN REQUIRE ONE OR MORE INTERMEDIATE HOSTS, EXCEPT:
   A. Taenia solium
   B. Taenia saginata
   C. Hymenolepis nana
   D. Diphyllobothrium latum
   E. Echinococcus granulosus
   Ref. 1 - p. 172

1495. THE USUAL SITES FOR THE HUMAN TAPEWORMS:
   A. Gallbladder, pancreatic duct
   B. Ileum, jejunum
   C. Pylorus, pancreatic duct
   D. Colon, sigmoidorectal region
   E. Appendix, cecum
   Ref. 1 - p. 171

1496. THE GRAVID PROGLOTTID OF TAENIA SAGINATA CAN BE BEST DIFFERENTIATED FROM THAT OF TAENIA SOLIUM BY THE:
   A. Morphology of the ova within the uterus
   B. Number of ova within the uterus
   C. Width of the proglottid
   D. Length of the proglottid
   E. Number of lateral uterine branches
   Ref. 1 - p. 187

1497. THE SECOND INTERMEDIATE HOST OF DIPHYLLOBOTHRIUM LATUM:
   A. Snail
   B. Fresh-water fish
   C. Copepod
   D. Water chestnut
   E. Salt-water fish
   Ref. 1 - p. 176

SECTION VI - MEDICAL PARASITOLOGY

1498. IN HYDATID DISEASE, MAN IS A "BLIND ALLEY" FOR THE PARASITE AS IN:
    A. Trichiniasis
    B. Enterobiasis
    C. Ascariasis
    D. Taeniasis
    E. None of the above
    Ref. 11 - p. 167

1499. IN ORDER TO PROCEED WITH THEIR DEVELOPMENT, THE EGGS OF HYMENOLEPIS DIMINUTA MUST BE INGESTED BY:
    A. A suitable arthropod
    B. A dog
    C. A cat
    D. All of the above.
    E. None of the above
    Ref. 9 - p. 186

1500. THE MOST COMMON INTERMEDIATE HOST FOR ECHINOCOCCUS GRANULOSUS:
    A. Man
    B. Dog
    C. Fox
    D. Wolf
    E. Sheep
    Ref. 1 - p. 193

ANSWER THE FOLLOWING QUESTIONS BY USING THE KEY OUTLINED BELOW:
    A. If only A is correct
    B. If only B is correct
    C. If both A and B are correct
    D. If neither A nor B is correct

1501. EXCEPT FOR SOME SPECIES OF SCHISTOSOMES, PARASITIC TREMATODES OF MAN ARE:
    A. Hermaphroditic
    B. Unisexual
    C. Both
    D. Neither
    Ref. 9 - p. 124

1502. PRIMARY INTERMEDIATE HOSTS FOR HUMAN PARASITIC TREMATODES:
    A. Appropriate molluscs
    B. Brackish-water fishes
    C. Both
    D. Neither
    Ref. 9 - p. 124

1503. THE SYSTEMIC REACTIONS PRODUCED BY FLUKES ARE USUALLY DUE TO:
    A. Intravascular hemolysis
    B. Absorption of toxic substances
    C. Both
    D. Neither
    Ref. 1 - p. 212

1504. CONCERNING FASCIOLOPSIS BUSKI:
    A. The largest parasitic trematode of man
    B. Inhabits the duodenum and jejunum
    C. Both
    D. Neither
    Ref. 1 - p. 213

1505. THE LIFE CYCLE OF CLONORCHIS SINENSIS INVOLVES:
    A. Snails
    B. Fresh-water fishes
    C. Both
    D. Neither
    Ref. 1 - pp. 223-224

1506. MAN CONTRACTS BEEF TAPEWORM INFECTION BY INGESTING RAW OR RARE BEEF CONTAINING:
    A. Cysticercus cellulosae
    B. Cysticercus bovis
    C. Both
    D. Neither
    Ref. 11 - p. 159

1507. THE INTESTINAL TRACT AND THE LIVER BEAR THE BRUNT OF THE DAMAGE IN INFECTIONS CAUSED BY:
A. Schistosoma japonicum
B. Schistosoma mansoni
C. Both
D. Neither
Ref. 11 - p. 181

1508. THE DISEASE CAUSED BY SCHISTOSOMA HAEMATOBIUM IS KNOWN AS:
A. Urinary bilharziasis
B. Vesical schistosomiasis
C. Both
D. Neither
Ref. 1 - p. 241

1509. THE SCHISTOSOMES DIFFER FROM OTHER TYPICAL TREMATODES IN HAVING:
A. Separate sexes
B. Narrow elongated shape
C. Both
D. Neither
Ref. 1 - p. 234

1510. THE INCUBATION PERIOD OF INTESTINAL SCHISTOSOMIASIS IS INITIATED BY THE:
A. Penetration of the skin by cercariae
B. Penetration of the skin by metacercariae
C. Both
D. Neither
Ref. 1 - p. 245

1511. IN MAN, OVA OF SCHISTOSOMA JAPONICUM ARE USUALLY FOUND IN THE:
A. Uterine plexuses
B. Urine
C. Both
D. Neither
Ref. 1 - pp. 245-246

1512. ENVENOMIZATION IS CAUSED BY:
A. Centipedes
B. Scorpions
C. Both
D. Neither
Ref. 11 - p. 223

1513. MAN IS AN IMPORTANT HOST OF:
A. Pulex irritans
B. Tunga penetrans
C. Both
D. Neither
Ref. 1 - p. 260

1514. THE LARVAL STAGE OF TROMBICULA AKAMUSHI TRANSMITS:
A. Scrub typhus
B. Epidemic typhus
C. Both
D. Neither
Ref. 11 - p. 253

1515. IMPORTANT REPRESENTATIVE OF THE ORDER HEMIPTERA:
A. Climex lectularius
B. Triatomid bugs
C. Both
D. Neither
Ref. 11 - p. 257

1516. SANDFLIES ARE THE VECTORS OF:
A. Leishmaniasis
B. Malaria
C. Both
D. Neither
Ref. 11 - p. 262

1517. AFRICAN TRYPANOSOMIASIS IS TRANSMITTED BY:
A. Tsetse flies
B. Sandflies
C. Both
D. Neither
Ref. 11 - p. 262

SECTION VI - MEDICAL PARASITOLOGY

1518. HARD TICKS TRANSMIT:
 A. American spotted fever C. Both
 B. Colorado tick fever D. Neither
 Ref. 11 - p. 250

1519. THE FOLLOWING SPECIMEN IS USUALLY SUBMITTED FOR EXAMINATION OF TOXOPLASMA GONDII:
 A. Sputum C. Both
 B. Biopsied tissues D. Neither
 Ref. 11 - p. 89

1520. DUODENAL CONTENTS MAY BE OF VALUE FOR:
 A. Giardia trophozoites C. Both
 B. Clonorchis eggs D. Neither
 Ref. 1 - p. 319

1521. LYMPH NODE BIOPSY MAY BE USED FOR THE DIAGNOSIS OF:
 A. Trypanosomiasis C. Both
 B. Leishmaniasis D. Neither
 Ref. 1 - p. 327

1522. KNOTT CONCENTRATION TECHNIC IS USED TO DETECT:
 A. Malarial parasites C. Both
 B. Microfilariae D. Neither
 Ref. 1 - p. 326

1523. SEROLOGICAL AND SKIN TESTS HAVE BEEN FOUND TO BE USEFUL IN:
 A. Trichinosis C. Both
 B. Echinococcosis D. Neither
 Ref. 1 - p. 316

1524. SCRAPINGS FROM ANAL AND PERIANAL REGIONS CAN BE USED FOR OVA COLLECTION OF:
 A. Enterobius vermicularis C. Both
 B. Taenia saginata D. Neither
 Ref. 1 - p. 320

1525. THE SABIN-FELDMAN DYE TEST IS USEFUL FOR THE DIAGNOSIS OF:
 A. Trypanosomiasis C. Both
 B. Toxoplasmosis D. Neither
 Ref. 1 - p. 317

ANSWER THE FOLLOWING QUESTIONS BY USING THE KEY OUTLINED BELOW:
 A. If 1 and 4 are correct
 B. If 2 and 3 are correct
 C. If 1, 2 and 3 are correct
 D. If all are correct
 E. If all are incorrect
 F. If some combination other than the above is correct

1526. THE TROPHOZOITES OF GIARDIA LAMBLIA PRODUCE IRRITATION BY ATTACHING TO THE WALL OF THE:
 1. Small intestine
 2. Common bile duct
 3. Gallbladder
 4. Large intestine Ref. 11 - p. 43

SECTION VI - MEDICAL PARASITOLOGY

1527. MALARIAL PARASITES REQUIRE FROM THE HOST FOR NOURISHMENT:
    1. Carbohydrates
    2. Proteins
    3. Fats
    4. Methionine   Ref. 1 - p. 78

1528. IN ACUTE VIVAX MALARIA INFECTIONS, THE LIVER IS:
    1. Soft
    2. Firm
    3. Enlarged
    4. Slightly enlarged   Ref. 1 - p. 85

1529. IN INFLAMMATORY FILARIASIS, THE LYMPHATICS IN THE FOLLOWING AREAS ARE CHIEFLY AFFECTED:
    1. Arms
    2. Legs
    3. Genitalia
    4. Neck   Ref. 1 - p. 147

1530. WHICH OF THE MALARIAL PARASITES HAS AN AFFINITY FOR RETICULOCYTES?:
    1. P. vivax
    2. P. malariae
    3. P. falciparum
    4. P. ovale   Ref. 1 - p. 76

1531. THE PATIENT WITH INTESTINAL AMEBIASIS WILL USUALLY HAVE:
    1. Tenesmus
    2. Abdominal pain
    3. Abdominal tenderness
    4. Acute systemic intoxication   Ref. 9 - p. 74

1532. FALCIPARUM INFECTION DIFFERS FROM OTHER TYPES IN THAT:
    1. The schizogonic cycle is frequently less synchronized
    2. There is a tendency for more than one parasite to develop in a single red cell
    3. Infected red cells tend to adhere to one another
    4. There is progressive increase in the number and quality of circulating erythrocytes   Ref. 9 - p. 104

1533. IN CONTRAST TO SCHISTOSOMIASIS JAPONICA, IN SCHISTOSOMIASIS MANSONI:
    1. Hepatic cirrhosis develops much more gradually
    2. Frank ascites is more frequent
    3. Fibrosis of the mesentery-omentum is invariably demonstrated
    4. Early intestinal lesions develop typically in the colon
                                 Ref. 9 - pp. 164-165

1534. THE AVAILABLE METHODS OF CLINICAL LABORATORY DIAGNOSIS OF TRICHINOSIS INCLUDE:
    1. X-ray
    2. Muscle biopsy
    3. Immunologic tests
    4. Blood chemistry   Ref. 9 - p. 220

1535. THE TYPE OF ANEMIA PRODUCED BY HOOKWORMS IS TYPICALLY:
    1. Microcytic
    2. Macrocytic
    3. Hyperchromic
    4. Hypochromic   Ref. 9 - p. 248

SECTION VI - MEDICAL PARASITOLOGY

ANSWER THE FOLLOWING QUESTIONS BY USING THE KEY OUTLINED BELOW:
A. If both statement and reason are true and related cause and effect
B. If both statement and reason are true but not related cause and effect
C. If the statement is true but the reason is false
D. If the statement is false but the reason is true
E. If both statement and reason are false

1536. Blacks are more resistant than whites to vivax malaria BECAUSE active immunity is acquired naturally by previous infection.
Ref. 1 - p. 10

1537. Therapeutic reduction of human sources of parasitic infection is not a practical measure BECAUSE it is easier to eradicate the animal reservoir hosts by therapeutic measures.
Ref. 1 - p. 12

1538. Trophozoites of parasitic protozoa are extremely difficult to kill even in an unfavorable environment BECAUSE they are the actively growing forms.
Ref. 1 - p. 14

1539. Encystment of Entamoeba histolytica is essential for transmission BECAUSE only the mature trophozoite is infectious.
Ref. 1 - p. 22

1540. It is possible to differentiate the trophozoite of Entamoeba histolytica from that of E. coli BECAUSE the former has evenly distributed chromatin granules on the nuclear membrane with a centrally located karyosome.
Ref. 1 - p. 20

1541. The life cycle of Entamoeba histolytica is comparatively simple BECAUSE man is the principal host and source of infection.
Ref. 1 - p. 24

1542. Pulmonary amebiasis is a frequent clinical occurrence BECAUSE it usually results from direct extension of an hepatic abscess.
Ref. 1 - p. 29

1543. Trichomonas vaginalis is frequently found in urine BECAUSE its habitat is the vagina of the female and the urethra of the male.
Ref. 1 - p. 41

1544. Massive malaria infection of the placenta may occur without infection of the child BECAUSE the placenta acts as a barrier.
Ref. 1 - p. 84

1545. Treatment of Chagas' disease is unsatisfactory BECAUSE there is no drug that destroys the tissue-inhabiting organisms.
Ref. 1 - p. 58

1546. Toxoplasma gondii is an obligate intracellular parasite BECAUSE it has a cell membrane, a spheroid-to-oval nucleus with a central karyosome.
Ref. 1 - p. 68

1547. The Sabin-Feldman dye test is the most useful serologic test for all stages of toxoplasmosis BECAUSE the test becomes positive in 2 to 3 weeks after the infection.
Ref. 1 - p. 70

1548. In endemic regions, chronic malaria may become a serious debilitating disease BECAUSE of relapses, sequelae, and reinfections.
Ref. 1 - p. 89

SECTION VI - MEDICAL PARASITOLOGY

1549. About 25% of patients with malaria give false positive reactions with the Kahn test for syphilis BECAUSE malaria is more common in patients with syphilis. Ref. 1 - p. 97

1550. Malaria cannot be transmitted by blood transfusions BECAUSE malaria is normally transmitted by mosquitoes.
Ref. 1 - p. 84

1551. Serious and sometimes fatal effects of ascariasis can occur BECAUSE of the migrations of the adult worms.
Ref. 1 - p. 136

1552. The protean symptoms of trichinosis simulate those of many other diseases BECAUSE of the involvement of many organs.
Ref. 1 - p. 109

1553. Clinical trichuriasis cannot be differentiated from other intestinal nematode infections BECAUSE the whipworms produce a wide variety of harmless to serious effects. Ref. 1 - p. 116

1554. Hog is the principal host of Strongyloides stercoralis BECAUSE after leaving the host, the free-living rhabditiform larvae may develop into infective filariform larvae. Ref. 1 - p. 117

1555. The filariform larvae of hookworms gain access to the host through hair follicles BECAUSE light sandy soil facilitates infection.
Ref. 1 - p. 125

1556. Hookworm infection can be reduced or even eliminated BECAUSE the only preventive measure consists of treatment of infected individuals.
Ref. 1 - p. 128

1557. Stool examination is of little value in enterobiasis BECAUSE eggs are infrequently laid in the intestine. Ref. 1 - p. 131

1558. The development of Ascaris lumbricoides ova is retarded in putrefactive material BECAUSE the ova require oxygen.
Ref. 1 - p. 134

1559. Reactions are accelerated and intensified in individuals reinfected with Ascaris BECAUSE the individuals develop a striking generalized hypersensitivity to the larval metabolites. Ref. 11 - p. 107

1560. Nocturnal periodicity with Wuchereria bancrofti is absolute BECAUSE no microfilariae can be found in the blood during the day.
Ref. 1 - p. 148

1561. The intracutaneous test in toxoplasmosis has limited diagnostic value BECAUSE it develops late in the disease.
Ref. 1 - p. 72

1562. Definitive diagnosis of paragonimiasis can be diagnosed from the roentgenogram BECAUSE the chest shadows are quite different from those of pulmonary tuberculosis. Ref. 9 - p. 148

1563. Transfusion malaria infections are usually easily eradicated BECAUSE they reproduce only the erythrocytic cycle.
Ref. 1 - p. 84

SECTION VI - MEDICAL PARASITOLOGY

1564. The prevalence rate of human trichinosis in the U.S. is the lowest in the world BECAUSE the control of the disease has been extremely successful.
Ref. 11 - p. 327

1565. Enterobius vermicularis is never found in the appendix BECAUSE the usual habitat of the mature pinworm is the cecum.
Ref. 9 - p. 226

1566. Rectal biopsy is a reliable diagnostic procedure in schistosomiasis mansoni BECAUSE characteristic eggs are recovered only by this procedure.
Ref. 9 - p. 165

1567. The methylene blue dye test is used widely in the diagnosis of toxoplasmosis BECAUSE this test is dependent on the fact that in the presence of specific antibodies the cytoplasm of the organism loses its affinity for the dye.
Ref. 11 - p. 90

1568. Diagnosis of balantidiasis depends upon the identification of trophozoites in diarrheic stools BECAUSE cysts are not found in stools.
Ref. 1 - p. 37

1569. The main source of amebiasis is the cyst passer BECAUSE viable cysts have been recovered from the vomitus, feces, and flies.
Ref. 1 - p. 25

1570. Alveolar hydatid disease is almost invariably fatal BECAUSE the organism grows without capsular confinement.
Ref. 9 - p. 202

1571. Falciparum malaria is accompanied by pernicious manifestations more frequently than in the other types BECAUSE vivax and quartan malaria are relatively benign.
Ref. 9 - p. 105

1572. Schistosomiasis caused by S. japonicum is more severe than that caused by S. mansoni BECAUSE the female worm extrudes about ten times as many eggs as S. mansoni.
Ref. 1 - p. 246

1573. In general, pulmonary pathologic changes in human strongyloidiasis are not marked BECAUSE although the larvae are carried to the lungs via the blood stream, they do not migrate into the alveoli.
Ref. 9 - p. 255

1574. Younger children are particularly favorable subjects for infection with Hymenolepis nana BECAUSE the infection is essentially one of anus-to-mouth transmission.
Ref. 9 - p. 185

1575. Definitive diagnosis of malaria is made by the identification of the parasites in blood smears BECAUSE parasites may be absent between attacks, after treatment, or during suppressive therapy.
Ref. 1 - p. 90

SECTION VI - MEDICAL PARASITOLOGY

ANSWER THE FOLLOWING QUESTIONS BY USING THE KEY OUTLINED BELOW:
A. If A is greater in frequency and/or magnitude than B
B. If B is greater in frequency and/or magnitude than A
C. If A and B are approximately equal

1576. INCIDENCE OF GIARDIASIS BY AGE GROUP:
A. 10-year-old
B. 40-year-old   Ref. 1 - p. 40

1577. THE SEVERITY OF THE DISEASE:
A. Gambian trypanosomiasis
B. Rhodesian trypanosomiasis   Ref. 1 - p. 54

1578. FREQUENCY OF MIXED MALARIAL INFECTIONS:
A. P. falciparum and P. vivax
B. P. vivax and P. malariae   Ref. 1 - p. 76

1579. LENGTH OF LIFE SPAN IN MAN OF:
A. P. falciparum
B. P. malariae   Ref. 1 - p. 89

1580. THE SEVERITY OF THE PRIMARY ATTACK DUE TO:
A. Plasmodium malariae
B. Plasmodium falciparum   Ref. 1 - p. 88

1581. THE MOST COMMON FORM OF HYDATID CYST FOUND IN MAN:
A. Unilocular
B. Alveolar   Ref. 1 - p. 193

1582. THE INCIDENCE OF ENTEROBIUS INFECTION IN:
A. Adults
B. Children   Ref. 9 - p. 229

1583. THE LENGTH OF THE ADULT WORM:
A. Taenia saginata
B. Taenia solium   Ref. 1 - p. 6

1584. LENGTH OF BUCCAL CHAMBER OF RHABDITIFORM LARVAE:
A. Strongyloides stercoralis
B. Necator americanus   Ref. 11 - p. 211

1585. SIZE OF MICROFILARIA:
A. Brugia malayi
B. Wuchereria bancrofti Ref. 9 - p. 280

1586. DURATION OF SCHIZOGONY:
A. Plasmodium vivax
B. Plasmodium falciparum Ref. 9 - p. 108

1587. DURATION OF FEBRILE PAROXYSM IN PRIMARY ATTACK:
A. Plasmodium falciparum
B. Plasmodium vivax   1 - p. 87

1588. PERSISTENCE OF EXO-ERYTHROCYTIC FORMS IN:
A. Plasmodium falciparum
B. Plasmodium vivax   Ref. 11 - pp. 76-77

1589. APPROXIMATE SIZE OF EGG:
A. Fasciolopsis buski
B. Paragonimus westermani Ref. 11 - p. 211

1590. PATHOGENICITY OF:
A. Trichomonas hominis
B. Chilomastix mesnili   Ref. 11 - p. 45

1591. DURATION OF VIABILITY IN THE SOIL OF INFECTIVE EGGS:
A. Ascaris lumbricoides
B. Trichuris trichiura   Ref. 11 - pp. 103,107

1592. SIZE OF THE EGGS:
A. Taenia solium
B. Taenia saginata   Ref. 11 - p. 162

1593. EXTENSIVENESS OF EXTRAINTESTINAL PATHOLOGY RESULTING FROM:
A. Balantidium coli
B. Entamoeba histolytica   Ref. 11 - p. 57

1594. LIVER INVOLVEMENT IN INFECTIONS DUE TO:
A. Schistosoma japonincum
B. Schistosoma mansoni   Ref. 11 - p. 181

1595. NUMBER OF NUCLEI IN MATURE CYST:
A. Endolimax nana
B. Entamoeba histolytica   Ref. 11 - p. 200

1596. INCUBATION PERIOD:
A. Echinococcosis
B. Cysticercosis   Ref. 11 - p. 172

1597. DEGREE OF ANEMIA IN MALARIA DUE TO:
A. Plasmodium vivax
B. Plasmodium falciparum   Ref. 11 - p. 79

1598. APPROXIMATE SIZE OF EGGS:
A. Schistosoma mansoni
B. Ascaris lumbricoides   Ref. 11 - p. 211

1599. NUMBER OF MEROZOITES IN THE MATURE SCHIZONT:
A. Plasmodium vivax
B. Plasmodium malariae   Ref. 11 - p. 83

1600. PROGRESSIVE MOTILITY OF TROPHOZOITE:
A. Entamoeba coli
B. Entamoeba histolytica   Ref. 11 - p. 200

SELECT THE ITEM FROM COLUMN I WHICH IS UNRELATED TO THE FOUR OTHER ITEMS IN THIS COLUMN, CIRCLE THIS LETTER. IDENTIFY THE STATEMENT IN COLUMN II WITH WHICH THE FOUR RELATED ITEMS IN COLUMN I ARE ASSOCIATED, CIRCLE THIS NUMBER:

	I	II
1601.	A. Hymenolepis nana	1. One-cell stage egg in stool
	B. Schistosoma japonicum	2. 4-8-cell stage egg in stool
	C. Schistosoma mansoni	3. Embryonated egg in stool
	D. Clonorchis sinensis	
	E. Diphyllobothrium latum	Ref. 11 - p. 211

SECTION VI - MEDICAL PARASITOLOGY

1602.
A. Bancroft's filariasis
B. Onchocerciasis
C. Loaiasis
D. Acanthocheilonemiasis
E. Ozzard's filariasis

1. Mosquitoes serve as vectors
2. Insects other than mosquitoes serve as vectors
3. Small crustaceans serve as vectors
Ref. 11 - p. 146

1603.
A. Ascaris lumbricoides
B. Enterobius vermicularis
C. Trichuris trichiura
D. Trichinella spiralis
E. Necator americanus

1. Portal of entry in man: mouth
2. Portal of entry in man: skin
3. Portal of entry in man: genitalia
Ref. 1 - p. 4

1604.
A. Paragonimus westermani
B. Ascaris lumbricoides
C. Fasciola hepatica
D. Hookworms
E. Strongyloides stercoralis

1. Sputum specimens may reveal eggs or larvae
2. Skin biopsy useful in diagnosis
3. Lymph node biopsy useful in diagnosis
Ref. 1 - pp. 327-328

1605.
A. Centipedes
B. Spiders
C. Scorpions
D. Ticks
E. Mites

1. Envenomization in man
2. Dermatitis in man
3. Obligatory tissue injury in man
Ref. 11 - p. 245

1606.
A. Colorado tick fever
B. American spotted fever
C. Tularemia
D. Scrub typhus
E. Endemic relapsing fever

1. Trnasmitted by ticks
2. Transmitted by mites
3. Transmitted by fleas
Ref. 11 - p. 266

1607.
A. Malaria
B. Yellow fever
C. Dengue
D. Western equine encephalitis
E. St. Louis encephalitis

1. Anopheles is vector
2. Mosquitoes other than Anopheles are vectors
3. Insects other than mosquitoes are vectors
Ref. 11 - p. 268

1608.
A. Enterobius vermicularis
B. Hymenolepis nana
C. Diphyllobothrium latum
D. Taenia saginata
E. Taenia solium

1. Common intestinal nematodes
2. Common intestinal cestodes
3. Common intestinal trematodes
Ref. 11 - p. 206

1609.
A. Entamoeba histolytica
B. Entamoeba coli
C. Endolimax nana
D. Dientamoeba fragilis
E. Iodamoeba bütschlii

1. Diagnosis can be made from iodine-stained cysts
2. Diagnosis can be made from iodine-stained trophozoites
3. No cysts have been reported
Ref. 11 - p. 200

1610.
A. Fleas
B. Mites
C. Lice
D. Flies
E. Bugs

1. Belong to Class Insecta
2. Belong to Class Arachnida
3. Belong to Class Crustacea

Ref. 11 - p. 248

SECTION VI - MEDICAL PARASITOLOGY

I | II

1611.
A. Fasciolopsis buski
B. Clonorchis sinensis
C. Paragonimus westermani
D. Diphyllobothrium latum
E. Schistosoma mansoni

1. Operculated eggs in stool
2. Nonoperculated eggs in stool
3. Gravid proglottids in stool

Ref. 11 - pp. 172, 190

1612.
A. Dwarf tapeworm infection
B. Broad tapeworm infection
C. Beef tapeworm infection
D. Cysticercosis
E. Pork tapeworm infection

1. Anemia is a striking clinical feature
2. Toxemia is a striking clinical feature
3. Ocular disturbance is a striking clinical feature

Ref. 11 - p. 172

1613.
A. Flagellum
B. Undulating membrane
C. Spiral groove
D. Blepharoplast
E. Axostyle

1. Parts of Entamoeba histolytica
2. Parts of Trichomonas vaginalis
3. Parts of Chilomastix mesnili

Ref. 11 - p. 53

1614.
A. Echinococcus granulosus
B. Hymenolepis nana
C. Diphyllobothrium latum
D. Taenia saginata
E. Taenia solium

1. Tissue cestodes
2. Intestinal cestodes
3. Intestinal trematodes

Ref. 11 - p. 149

1615.
A. Leishmania donovani
B. Leishmania tropica
C. Leishmania braziliensis
D. Trypanosoma gambiense
E. Trypanosoma cruzi

1. Trypanosomal stage is part of life cycle
2. Leishmanial stage is part of life cycle
3. Crithidial stage is part of life cycle

Ref. 11 - p. 58

MATCH EACH OF THE FOLLOWING CHOICES WITH THE MOST APPROPRIATE STATEMENTS OR CHOICES. EACH ITEM MAY BE USED ONLY ONCE:

A. "Black widow"
B. Diphyllobothrium latum
C. Loa loa
D. Paragonimus westermani
E. Onchocerciasis
F. Chagas' disease

1616. ___ Diaptomus spp.
1617. ___ Crayfish
1618. ___ Latrodectus mactans
1619. ___ Triatomid bug
1620. ___ Simulium spp.

Ref. 9 - pp. 378, 378, 326, 379, 380

SECTION VI - MEDICAL PARASITOLOGY

A. Enterobius vermicularis
B. Hymenolepis nana
C. Toxoplasma gondii
D. Trichuris trichiura
E. Taenia solium
F. Echinococcus granulosus

1621. ___ Whipworm
1622. ___ Dwarf tapeworm infection
1623. ___ Unilocular hydatid disease
1624. ___ Chorioretinitis   Ref. 11 - pp. 103, 150, 165, 89,
1625. ___ Pinworm   99

A. Trichinella spiralis
B. Fasciolopsis buski
C. Leishmania braziliensis
D. Schistosoma mansoni
E. Fasciola hepatica
F. Echinococcus granulosus

1626. ___ Man is intermediate host; dogs, wolves, and foxes are definitive hosts
1627. ___ Segmentina snail serves as first intermediate host; water caltrop and water chestnut serve as "second intermediate hosts"
1628. ___ Australorbis snail serves as intermediate host; man is invaded by forked-tail cercariae
1629. ___ The same animal acts as final and intermediate hosts, harboring the adult parasite temporarily and the larva for a longer period
1630. ___ Phlebotomus intermedius serves as the chief insect vector
                                  Ref. 1 - pp. 192-193, 213-214,
                                               245, 105, 67

A. Schistosoma japonicum
B. Wuchereria bancrofti
C. Schistosoma mansoni
D. Necator americanus
E. Hymenolepis nana
F. Schistosoma haematobium

1631. ___ Egg with "hook-cavity"
1632. ___ Sheathed microfilaria
1633. ___ Egg with large lateral spine
1634. ___ Egg with "polar filaments"   Ref. 11 - pp. 211, 142, 211, 152,
1635. ___ Semilunar cutting plates   116

AFTER EACH ONE OF THE FOLLOWING CASE HISTORIES ARE
SEVERAL MULTIPLE CHOICE QUESTIONS BASED ON THE HISTORY.
ANSWER BY CHOOSING THE MOST APPROPRIATE ANSWER:

CASE HISTORY - Questions 1636-1640

A white, unmarried male, age fifty, veteran of World War II, has had trouble with "bowels" for more than 20 years. He had been hospitalized for "dysentery" in November, 1944, to January, 1945. He has had one or two days of diarrhea each week with constipation between attacks of diarrhea. Relevant findings on admission to the hospital: chronically ill, malnourished appearance, pale, and thin with a striking tenderness over entire abdomen. Liver was not enlarged, but was tender on pressure. Laboratory findings were: Hb, 12.5 gm; RBC, 3.85; WBC, 7,600. Parasitology diagnostic laboratory reported protozoan trophozoites and some cysts in "mushy" specimen.

1636. THE MOST LIKELY DIAGNOSIS IS:
A. Amebic liver abscess
B. Intestinal amebiasis
C. Balantidiasis
D. Giardiasis
E. None of the above
Ref. 11 - p. 32

1637. IF DIAGNOSIS IS CORRECT, THE MOTILITY OF THE TROPHOZOITE SHOULD BE:
A. Progressive with an explosive extrusion of ectoplasmic pseudopodia
B. Sluggish and rarely progressive with pseudopodia extruded slowly
C. Active and progressive with pseudopodia extruded slowly
D. Active with indefinite progression and pseudopodia extruded slowly
E. None of the above
Ref. 11 - p. 200

1638. IN THE PATHOLOGY OF INTESTINAL AMEBIASIS:
A. The most common site for lesions is the cecal area
B. The lesion spreads to develop a flask-shaped lesion in the mucosa
C. In certain cases, trophozoites erode the muscularis mucosae and spread out radially in the surrounding submucosa
D. The trophozoites may metastasize to extraintestinal sites
E. All of the above
Ref. 11 - pp. 23-24

1639. THE MOST COMMON EXTRAINTESTINAL SITE FOR AMEBIASIS:
A. Lung
B. Brain
C. Liver
D. Spleen
E. Skin
Ref. 11 - p. 25

1640. COMMON MODE OF TRANSMISSION OF E. HISTOLYTICA CYSTS:
A. Polluted water supply
B. Infected food handlers
C. Mechanical transmission by flies
D. All of the above
E. None of the above
Ref. 11 - pp. 312-313

SECTION VI - MEDICAL PARASITOLOGY

CASE HISTORY - Questions 1641-1645

A child of 18 months was seen whose development had been essentially normal until the age of 12 months but appetite was not good and had intermittent diarrhea. She became ill two months ago, and lost weight rapidly. One month ago, watery diarrhea was noted; stools were yellow or green. She developed a loose, nonproductive "barking" cough. She was pale, and fretful when disturbed. A few crusted healing lesions (0.5 x 0.5 cm) were seen on the left foot. She appeared severely malnourished and dehydrated. Abdomen was distended; and bronchial fremitus palpable throughout both lungs. Hemoglobin was 9 gm.; RBC 3.92; and WBC 24,000. Differential showed 36 per cent eosinophils.

1641. THE MOST LIKELY DIAGNOSIS IS:
A. Strongyloidiasis
B. Trichinosis
C. Trichuriasis
D. Ascariasis
E. Enterobiasis   Ref. 11 - p. 133

1642. IF IT IS STRONGYLOIDIASIS, THE FOLLOWING STAGE OF THE PARASITE WILL BE FOUND IN THE FECAL SPECIMEN:
A. Filariform larvae
B. Rhabditiform larvae
C. One-cell stage eggs
D. 4-8-cell stage eggs
E. Many adult worms   Ref. 11 - pp. 126, 133

1643. THE PROMINENT LESIONS ON THE SOLE OF THE LEFT FOOT RESEMBLE:
A. Ascariasis
B. Trichuriasis
C. Hookworm infection
D. Trichiniasis
E. Oxyuriasis   Ref. 11 - p. 125

1644. THE DIAGNOSTIC FEATURE OF STRONGYLOIDES LARVAE AS COMPARED TO HOOKWORM LARVAE:
A. The rhabditiform larva has a short buccal cavity
B. The rhabditiform larva has a conspicuous genital primordium
C. The filariform larva has a characteristic notched tail
D. All of the above
E. None of the above   Ref. 11 - p. 126

1645. OCCASIONALLY, STRONGYLOIDES LARVAE ARE RECOVERED FROM SPECIMENS OTHER THAN FECES:
A. Sputum
B. Urine
C. Duodenal aspirates
D. All of the above
E. None of the above   Ref. 11 - p. 126

SECTION VI - MEDICAL PARASITOLOGY

CASE HISTORY - Questions 1646-1650

A black female, age five, was admitted because she had been ill for 2 days with abdominal pain and frequent retching. Blood had been noticed in the stools by the child's sister. The child was known to play in shaded areas close to the house. Temperature was 99.5°F; pulse, 90. Abdomen was moderately distended, with tenderness and slight rigidity in the right iliac fossa. On deeper palpation intestinal coils containing longitudinal masses, like bundles of pencils, were felt through the thin abdominal wall.

1646. THE HISTORY AND PHYSICAL FINDINGS SUGGEST:
A. Hookworm infection
B. Beef tapeworm infection
C. Schistosomiasis
D. Ascariasis
E. Trichuriasis   Ref. 11 - p. 111

1647. THE ADULT WORMS OF THIS PARASITE USUALLY LIVE:
A. Unattached in the lumen of the small intestine
B. Attached to the mucosa of the small intestine
C. Attached to the mucosa of the cecum and appendix
D. Attached superficially to the mucosa of the cecum
E. Embedded in the deeper portions of the mucosa in the duodenal and jejunal areas   Ref. 11 - p. 107

1648. WHEN THE LARVAE MIGRATE THROUGH THE LUNGS, THERE IS:
A. Intense cellular reaction around the larvae
B. The serocellular exudate that results is composed chiefly of eosinophils
C. Lobular consolidation
D. Development of generalized hypersensitivity
E. All of the above   Ref. 11 - p. 107

1649. THE CHARACTERISTIC ASCARIS EGGS ARE:
A. In the 4-8-cell stage when discharged in the stool
B. Bluntly ovoid, 45 to 70μ x 35 to 50μ with a mammillated outer coating
C. Barrel-shaped with a transparent plug at each pole
D. Ovoid, 50 x 30μ, with a thin shell
E. Embryonated when discharged in the stool
     Ref. 11 - p. 108

1650. THE DRUG OF CHOICE FOR ELIMINATING THE ADULT ASCARIS:
A. Piperazine citrate
B. Hexylresorcinol
C. Hetrazan
D. Dithiazanine
E. Medicinal gentian violet   Ref. 11 - p. 109

SECTION VII - GENERAL REVIEW OF INFECTIOUS DISEASES

FOR EACH OF THE FOLLOWING MULTIPLE CHOICE QUESTIONS CHOOSE THE ONE MOST APPROPRIATE ANSWER:

1651. IN DIPHTHERIA:
A. The classic picture is always present
B. The pseudomembrane is pathognomonic
C. Both peripheral and cranial nerves are sensitive to the toxin
D. Complete obstruction of air passage does not occur
E. The effect of the toxin on the heart is apparent from the onset of the disease   Ref. 6 - p. 433

1652. IF A WOMAN HAS MEASLES, HER INFANT WILL BE IMMUNE TO THE DISEASE FOR THE FIRST:
A. 14 days
B. 100 days
C. 6-9 months
D. 1 year
E. 5 years
Ref. 4 - p. 795

1653. IN LEPROSY:
A. The organisms are not typically acid-fast
B. The organisms are rarely found in the lepromatous form
C. The organisms have now been grown on artificial media with certainty
D. The organisms are often found within the endothelial cells of blood vessels
E. The organisms are never found in mononuclear cells
Ref. 15 - p. 193

1654. THE KEY TO THE PREVENTION OF DIPHTHERIA IS:
A. Treatment with penicillin
B. Active immunization
C. Periodic Schick tests
D. Antitoxin injection
E. Toxin-antitoxin injection   Ref. 6 - p. 435

1655. THE EARLIEST SIGN FOLLOWING INFECTION WITH MEASLES VIRUS IS USUALLY:
A. Cough, fever, coryza
B. Rash
C. Bronchitis
D. Conjunctivitis
E. Koplik's spots
Ref. 4 - p. 792

1656. WHICH IS A RARE COMPLICATION OF MEASLES?:
A. Bronchopneumonia
B. Encephalitis
C. Otitis media and mastoiditis
D. Laryngitis
E. Myocarditis
Ref. 4 - p. 793

1657. THE PERIOD BETWEEN EXPOSURE AND APPEARANCE OF THE EXANTHEM IN RUBELLA IS:
A. 5 to 7 days
B. 10 to 12 days
C. 12 to 23 days
D. 20 to 30 days
E. 1 month
Ref. 4 - p. 805

1658. THE MAJOR PERIOD OF MEASLES VIRUS DISSEMINATION:
A. At the time of exposure
B. Before the rash appears
C. At the time the rash appears
D. After the rash disappears
E. None of the above
Ref. 4 - p. 795

1659. STRAWBERRY TONGUE AND EXUDATIVE TONSILLITIS ARE DIAGNOSTIC FEATURES OF:
A. Rubeola
B. Scarlet fever
C. Rubella
D. Varicella
E. Variola
Ref. 15 - p. 263

SECTION VII - GENERAL REVIEW OF INFECTIOUS DISEASES

1660. THE SPASMODIC STAGE OF PERTUSSIS:
   A. Begins with coryza, sneezing, and cough
   B. Is characterized by a violent, repetitive type of cough which forces the air out of the lungs and is followed by a sudden inspiratory "whoop"
   C. Lasts for less than a week
   D. The bronchial epithelium is normal during this stage
   E. There is no sign of hypersensitivity to the proteins of the bacteria during this stage   Ref. 6 - p. 425

1661. A REASONABLE INDEX OF AN INDIVIDUAL'S IMMUNITY TO MUMPS CAN BE OBTAINED BY A DETERMINATION OF:
   A. His hypensensitivity to the skin test antigen
   B. The level of the complement-fixing antibody
   C. The level of the antihemagglutinating antibody
   D. All of the above
   E. None of the above   Ref. 4 - p. 758

1662. WHAT PERCENTAGE OF SUSCEPTIBLES IS LIKELY TO BECOME INFECTED AFTER A SINGLE EXPOSURE TO A PATIENT IN THE EARLY STAGES OF VARICELLA?:
   A. Less than 10 per cent
   B. About 10 per cent
   C. About 25 per cent
   D. About 50 per cent
   E. About 70 per cent
   Ref. 4 - p. 922

1663. IT IS GENERALLY ACCEPTED THAT THE DANGER OF INFECTIVITY OF CHICKENPOX IS OVER:
   A. When the skin rash appears
   B. When all lesions are in the vesicular stage
   C. After the appearance of the final crop of vesicles
   D. After the rash becomes confluent
   E. After the pock develops a pellicle
   Ref. 4 - p. 922

1664. IF A SUSCEPTIBLE INDIVIDUAL IS VACCINATED WITHIN 1-3 DAYS FOLLOWING EXPOSURE TO SMALLPOX:
   A. Protection can be expected except in rare cases
   B. A modified attack of smallpox will develop
   C. It is too late and no protection can be expected
   D. Generalized vaccinia will develop
   E. He will give an accelerated reaction
   Ref. 4 - pp. 949-950

1665. IN ATYPICAL PNEUMONIA DUE TO MYCOPLASMA PNEUMONIAE:
   A. The incubation period is usually 2-3 weeks
   B. Illness develops gradually over a few days
   C. Other organ systems in addition to the respiratory tract may be affected during the infection
   D. All of the above
   E. None of the above   Ref. 6 - p. 631

1666. WHEN A PATIENT WITH MENINGITIS IS LYING ON HIS BACK AND A LEG IS FLEXED AT RIGHT ANGLES AT BOTH THE HIP AND THE KNEE, THE HAMSTRING SPASM PREVENTS THE LEG BEING STRAIGHTENED AT THE KNEE, IF THE HIP IS STILL MAINTAINED AT THE RIGHT ANGLE. THIS IS:
   A. Brudzinski's sign
   B. MacEwen's sign
   C. Myerson's sign
   D. Kernig's sign
   E. None of the above   Ref. 5 - p. 444

SECTION VII - GENERAL REVIEW
OF INFECTIOUS DISEASES

1667. BACTERIAL MENINGITIS DUE TO THE FOLLOWING ORGANISM
OCCURS MOST FREQUENTLY IN CHILDREN:
A. N. meningitidis
B. Streptococcus pyogenes
C. H. influenzae
D. D. pneumoniae
E. M. tuberculosis
Ref. 6 - p. 418

1668. WHICH STATEMENT DOES NOT APPLY TO THE "MINOR ILLNESS"
SYNDROME OF POLIOMYELITIS?:
A. Follows a few days after exposure
B. Coincides with the period of viremia
C. Coincides with the appearance of the virus in the throat and
intestinal excreta
D. The symptoms include listlessness, low-grade fever, headache,
sore throat and vomiting
E. Stiffness of the neck and the back
Ref. 4 - p. 446

1669. IN GENERAL, CASES OF POLIOMYELITIS ARE DEFINED AS
PARALYTIC ONLY IF MUSCLE WEAKNESS PERSISTS:
A. Beyond 1 to 2 days
B. Beyond 3 to 5 days
C. Beyond 1 week
D. Beyond 2 to 3 weeks
E. None of the above
Ref. 4 - p. 447

1670. THE MOST COMMON MANIFESTATION OF INVOLVEMENT OF THE
CENTRAL NERVOUS SYSTEM BY AN ECHOVIRUS:
A. Paralytic disease
B. Aseptic meningitis
C. Encephalitis
D. Epidemic myalgia
E. Ocular disturbances
Ref. 4 - p. 528

1671. ALUM-PRECIPITATED DIPHTHERIA TOXOID IS PREFERABLE TO
FLUID DIPHTHERIA TOXOID BECAUSE:
A. It produces longer duration of passive immunity
B. It provides a prolonged antigenic stimulus
C. It is absorbed more rapidly
D. It has a lower incidence of side reactions
E. None of the above
Ref. 2 - p. 658

1672. ERYTHROGENIC TOXIN CAPABLE OF CAUSING SCARLET FEVER IS
PRODUCED BY:
A. Majority of Group A streptococci strains
B. A small number of strains of Group A streptococci
C. A few strains of pneumococci
D. Approximately 50% of meningococci
E. None of the above
Ref. 5 - p. 367

1673. BORDETELLA PERTUSSIS:
A. Causes a disease which simulates diphtheria
B. Is gram-positive
C. Newborn has high antibody titer for pertussis
D. Is an obligate parasite, which survives only a short time outside
the human host
E. Can be easily distinguished morphologically from Bordetella
parapertussis
Ref. 5 - p. 744

1674. MUMPS VIRUS MAY CAUSE INFLAMMATION OF THE:
A. Pancreas
B. Ovaries
C. Sublingual glands
D. Brain
E. All of the above
Ref. 4 - pp. 759-760

SECTION VII - GENERAL REVIEW OF INFECTIOUS DISEASES

1675. SMALLPOX VACCINATION OF A NON-IMMUNE PERSON SHOULD RESULT IN:
   A. Maximal local reaction in 3-4 days
   B. Maximal local reaction in 8-10 days
   C. Generalized vaccinia
   D. Local red papule in 24 hours
   E. No local reaction   Ref. 4 - p. 948

ANSWER THE FOLLOWING QUESTIONS BY USING THE KEY OUTLINED BELOW:
   A. If 1 and 4 are correct
   B. If 2 and 3 are correct
   C. If 1, 2 and 3 are correct
   D. If all are correct
   E. If all are incorrect
   F. If some combination other than the above is correct

1676. IN RUBELLA:
   1. The rash usually appears 18 days after infection
   2. The virus can be isolated from nasopharyngeal secretions as early as 7 days before the appearance of the exanthema
   3. The virus can be isolated from the blood during the prodromal illness and for 1 to 2 days after the rash appears
   4. The pathological lesions are characteristic for the disease
                                        Ref. 12 - pp. 1370-1371

1677. PERTUSSIS OCCURS THROUGHOUT THE YEAR BUT THE NORTHERN STATES HAVE THE HIGHEST INCIDENCE IN:
   1. January
   2. May
   3. August
   4. February   Ref. 6 - p. 425

1678. HISTOPATHOLOGIC CHANGES IN THE HOST CELL AS A RESULT OF VIRAL INFECTION INCLUDE:
   1. Nucleolar displacement
   2. Margination of nuclear chromatin
   3. Presence of pathognomonic intracytoplasmic inclusions
   4. Presence of pathognomonic intranuclear inclusions
                                        Ref. 4 - p. 343

1679. ANTIBIOTICS THAT AFFECT THE BACTERIAL CELL WALL INCLUDE:
   1. Cycloserine
   2. Vancomycin
   3. Bacitracin
   4. Chloramphenicol   Ref. 6 - p. 174

1680. PENICILLIN-RESISTANCE FREQUENTLY IS DEMONSTRATED BY:
   1. Staphylococci
   2. Gonococci
   3. Streptococci
   4. T. pallidum   Ref. 6 - p. 177

1681. THE ANTIBIOTICS THAT ARE EFFECTIVE AGAINST CERTAIN OF THE PATHOGENIC FUNGI:
   1. Amphotericin B
   2. Cycloserine
   3. Polymyxin
   4. Nystatin   Ref. 6 - p. 179

SECTION VII - GENERAL REVIEW OF INFECTIOUS DISEASES

1682. AS A RESULT OF IMMUNIZATION OF INFANTS AND PRESCHOOL CHILDREN WITH DIPHTHERIA TOXOID:
   1. The incidence of diphtheria in children has been reduced
   2. The chances of reinforcing the immunity by subclinical infections have been reduced
   3. More adults have positive Schick tests
   4. Higher percentage of newborns are immune during the first months of life Ref. 6 - p. 434

1683. SERIOUS COMPLICATIONS OF SCARLET FEVER INCLUDE:
   1. Acute hemorrhagic glomerulonephritis
   2. Puerperal sepsis
   3. Subacute bacterial endocarditis
   4. Acute rheumatic fever Ref. 6 - p. 380

1684. MATERNAL INFECTION WITH RUBELLA IN THE FIRST TRIMESTER PERIOD OF PREGNANCY MAY CAUSE:
   1. Patent ductus arteriosus
   2. Congenital cataracts
   3. Interventricular septal defect
   4. Renal agenesis Ref. 4 - p. 805

1685. THE HERPES SIMPLEX VIRUS CONSISTS OF:
   1. DNA
   2. Protein
   3. Lipid
   4. Carbohydrate Ref. 4 - p. 893

1686. CONCERNING HERPES ZOSTER:
   1. Incubation period is uncertain
   2. The eruption is usually bilateral
   3. The pain is very characteristic as to timing, duration and extent
   4. There is early appearance of regional lymphadenopathy
      Ref. 6 - pp. 865-866

1687. WHICH IS CHARACTERISTIC OF CHICKENPOX?:
   1. The rash usually becomes confluent after a few days
   2. Centripetal distribution of rash
   3. All stages of lesions are found at the same time in the same vicinity
   4. Lesions are usually all in the same stage at the same hour
      Ref. 4 - p. 919

1688. THE RASH IN SMALLPOX:
   1. The focal rash usually appears first on the buccal and the pharyngeal mucosa
   2. Lesions in any one area are all at the same stage of development
   3. Crusting is usually complete in 14 to 16 days from the onset of the illness
   4. Lesions are most numerous in the groin and the axillae
      Ref. 4 - p. 939

1689. THE TYPES OF SMALLPOX WITH FREQUENT FATALITIES:
   1. Confluent smallpox
   2. Alastrim
   3. Varioloid
   4. Purpuric smallpox Ref. 4 - pp. 939-940

SECTION VII - GENERAL REVIEW OF INFECTIOUS DISEASES

1690. THE FOLLOWING BIRDS HAVE BEEN FOUND TO SERVE AS SOURCES FOR PSITTACOSIS IN MAN:
1. Parakeets
2. Ducks
3. Turkeys
4. Pigeons
Ref. 12 - p. 955

1691. VACCINATION OF A PREVIOUSLY VACCINATED PERSON WHO HAS NEVER HAD SMALLPOX WILL MANIFEST ITSELF AS:
1. An accelerated reaction
2. No reaction
3. Primary reaction
4. Vaccinoid
Ref. 4 - p. 948

1692. PARACOCCIDIOIDOMYCOSIS IS A CHRONIC GRANULOMATOUS INFECTION OF THE:
1. Mucous membranes of the mouth
2. Skin
3. Lymph nodes
4. Internal organs
Ref. 6 - p. 968

1693. RESPIRATORY FAILURE OCCURS IN SPINAL POLIOMYELITIS AS A RESULT OF:
1. Paralysis of the intercostals
2. Paralysis of the diaphragm
3. Paralysis of the abdominal muscles
4. Paralysis of the pharyngeal muscles
Ref. 4 - p. 448

1694. POLIOMYELITIS VIRUS IS PRESENT:
1. In the throat during the acute phase of illness
2. In fecal material for several weeks
3. In the fecal specimens in 80 to 90 per cent of cases in the first 10 days
4. In the blood during acute phase of illness
Ref. 4 - p. 449

1695. TYPHOID FEVER:
1. Incubation period is usually 14 days
2. Bacteremia probably occurs in the first two weeks of the disease
3. Rose spots appear in the skin of the abdomen between the 10th and 15th day of disease
4. Agglutinins do not reach levels of diagnostic significance until the end of the second or third week
Ref. 6 - p. 512

1696. CHRONIC PROGRESSIVE HISTOPLASMOSIS CLOSELY RESEMBLES:
1. The reinfection type of tuberculosis
2. Lepromatous form of leprosy
3. Chronic malaria
4. Secondary syphilis
Ref. 6 - p. 977

1697. IN RUBELLA:
1. The infection is via the respiratory tract with spread to lymphatic tissue but not to the blood
2. Both viremia and respiratory-tract shedding of virus may precede the rash
3. There is major excretion of virus prior to recognition of illness
4. Actual pathology of the postnatal disease is well known because of the high fatality rate
Ref. 6 - p. 945

SECTION VII - GENERAL REVIEW
OF INFECTIOUS DISEASES

1698. THE DYSENTERY BACILLI:
1. Appear in profusion in the stools
2. Appear in profusion in the ulcers of the intestinal tract
3. Rarely invade the blood stream
4. Appear in large numbers in the blood stream
Ref. 6 - p. 525

1699. ANTHRAX IN MAN:
1. Is acquired exclusively by accidental cutaneous inoculation
2. The organism or spores may be inhaled or ingested
3. Infection occurs most frequently on the hands and forearms
4. Prognosis is poor even if treated early
Ref. 6 - p. 584

1700. INFECTIOUS HEPATITIS:
1. Fever is common
2. HB Ag (Au antigen)
3. Occurs in autumn and winter
4. Is transmitted parenterally   Ref. 15 - p. 374

MATCH EACH OF THE FIVE ITEMS LISTED BELOW WITH THE MOST APPROPRIATE STATEMENTS OR CHOICES. EACH ITEM MAY BE USED ONLY ONCE:

A. Causes cataracts in young fetus
B. Glycerinated calf lymph vaccine
C. LEP Flury vaccine
D. Orchitis
E. Pannus

1701. ___ Rabies
1702. ___ Rubella
1703. ___ Smallpox
1704. ___ Trachoma
1705. ___ Mumps   Ref. 4 - pp. 822, 805, 950,
                              1049, 760

A. Rickettsia prowazeki
B. Rickettsia tsutsugamushi
C. Rickettsia mooseri
D. Rickettsia quintana
E. Rickettsia rickettsii

1706. ___ Flea typhus
1707. ___ Tick fever
1708. ___ Famine fever
1709. ___ Trench fever
1710. ___ Tropical typhus   Ref. 4 - pp. 1082, 1095, 1059,
                                    1161, 1130

A. Mantoux test
B. Ascoli test
C. Schultz-Charlton reaction
D. Schick reaction
E. Quellung reaction

1711. ___ Diphtheria
1712. ___ Scarlet fever
1713. ___ Tuberculosis
1714. ___ Pneumococci   Ref. 6 - pp. 433, 380, 459
1715. ___ Anthrax   Ref. 15 - pp. 173, 179

SECTION VII - GENERAL REVIEW OF INFECTIOUS DISEASES

  A. Coagulase
  B. Moloney test
  C. Streptolysin-S
  D. OT
  E. Oxidase

1716. ___ Streptococcus
1717. ___ Mycobacteria
1718. ___ Staphylococcus aureus
1719. ___ Corynebacterium diphtheriae   Ref. 6 - pp. 375, 459, 391, 435,
1720. ___ Neisseria   405

  A. Clostridium novyi
  B. Clostridium tetani
  C. Clostridium perfringens
  D. Clostridium butyricum
  E. Clostridium histolyticum

1721. ___ Gelatin liquefaction: + ; Milk: stormy fermentation
1722. ___ Gelatin liquefaction: + ; Milk: acid
1723. ___ Gelatin liquefaction: + ; Milk: digested
1724. ___ Gelatin liquefaction: + ; Milk: variable
1725. ___ Gelatin liquefaction: - ; Milk: stormy fermentation
                                         Ref. 6 - p. 591

  A. Plasmodium falciparum
  B. Trypanosoma cruzi
  C. Trichuris trichiura
  D. Schistosoma mansoni
  E. Enterobius vermicularis

1726. ___ Anal pruritus
1727. ___ Blackwater fever
1728. ___ Romaña's sign
1729. ___ Prolapse of rectum
1730. ___ Periportal fibrosis   Ref. 6 - pp. 1047, 1040, 1037,
                                                        1049, 1970

ANSWER THE FOLLOWING QUESTIONS BY USING THE KEY OUTLINED BELOW:
  A. If both statement and reason are true and related cause and effect
  B. If both statement and reason are true but not related cause and effect
  C. If the statement is true but the reason is false
  D. If the statement is false but the reason is true
  E. If both statement and reason are false

1731. Individuals who have been infected but never had a clinically recogniz-able case of the disease will be immune to the disease in question BECAUSE the inapparent subclinical infections will produce effective immunity, especially if repeated.   Ref. 10 - p. 555

1732. The serum of agammaglobulinemic individuals contains no gamma globulin or a very small amount BECAUSE such persons are found to be unable to produce antibodies.
                                         Ref. 10 - p. 100

SECTION VII - GENERAL REVIEW OF INFECTIOUS DISEASES

1733. Lymphogranuloma venereum occurs only in man under natural conditions BECAUSE it is transmitted by sexual intercourse.
Ref. 12 - p. 956

1734. Immunity to measles is not transmitted via the placenta to the fetus BECAUSE the attack rate is 90 per cent or higher among susceptibles who come in close contact with an early case.
Ref. 4 - p. 795

1735. Measles is one of the most highly communicable of all the viral infections BECAUSE the virus can be recovered from the blood, particularly from the white cell fraction, for several days before the rash appears.
Ref. 6 - p. 907

1736. Repeated attacks of rubella are frequent BECAUSE the immunity produced by the disease is short-lived. Ref. 4 - p. 807

1737. Pertussis vaccine should not be given to children three months old BECAUSE children of this age do not produce antibodies as readily as children nine months old. Ref. 6 - p. 426

1738. Sterility resulting from mumps orchitis is rare BECAUSE atrophy of all glandular tissue may not ensue, even in the severe cases.
Ref. 4 - p. 760

1739. It is thought that there is an etiological relationship between chickenpox and herpes zoster BECAUSE some cases of herpes zoster are the cause of chickenpox and there are instances of apparent cross protection. Ref. 4 - p. 915

1740. Clinical diagnosis of trichiniasis is very difficult BECAUSE human trichiniasis may resemble as many as 40 different diseases.
Ref. 11 - p. 219

1741. Geotrichosis must be differentiated from tuberculosis BECAUSE pulmonary geotrichosis simulates tuberculosis, with an elevation of temperature, pulse and respiration, and leukocytosis.
Ref. 8 - pp. 366, 374

1742. Malaria due to P. falciparum does not fit the classical clinical picture of malaria BECAUSE the blockage of vessels is most characteristic of falciparum malaria. Ref. 11 - pp. 80, 81

1743. Extensive dissemination of enteroviruses in a population can interfere with the effectiveness of oral poliovirus vaccines BECAUSE oral poliovirus vaccines induce actual infection.
Ref. 4 - p. 460

1744. Tonsillectomy should not be performed during an epidemic of poliomyelitis BECAUSE it probably favors the occurrence of the bulbar form of the disease. Ref. 5 - p. 445

1745. Disseminated sporotrichosis is relatively rare BECAUSE cutaneous lymphatic sporotrichosis is the most common form of the disease.
Ref. 8 - pp. 419, 423

SECTION VII - GENERAL REVIEW OF INFECTIOUS DISEASES

1746. The virus cannot be isolated from the blood at the time of the onset of clinical symptoms of mumps BECAUSE mumps has an acute onset of fever and inflammation of salivary glands.
Ref. 12 - p. 1355

1747. Hemorrhoids due to portal obstruction may be the first indication of schistosomiasis mansoni BECAUSE the main damage and symptoms result from the spined eggs. Ref. 11 - p. 181

1748. The incubation period of cholera is short BECAUSE there is tremendous loss of fluid, dehydration, and metabolic acidosis.
Ref. 2 - p. 537

1749. Ornithosis was originally called psittacosis BECAUSE the disease is acquired exclusively from pisttacine birds.
Ref. 2 - p. 850

1750. Lymphogranuloma venerum is known as tropical bubo BECAUSE the disease occurs only in tropical regions, and not in temperate zones.
Ref. 2 - p. 853

ANSWER THE FOLLOWING QUESTIONS BY USING THE KEY OUTLINED BELOW:
A. If only A is correct
B. If only B is correct
C. If both A and B are correct
D. If neither A nor B is correct

1751. MAN IS INTERMEDIATE HOST FOR:
A. Cysticercus cellulose
B. Cysticercus bovis
C. Both
D. Neither
Ref. 11 - p. 172

1752. THE EFFECTIVENESS OF THE DRUG IN THE TREATMENT OF TYPHOID FEVER:
A. Penicillin
B. Chlortetracycline
C. Both
D. Neither
Ref. 6 - p. 513

1753. AN ATMOSPHERE OF 10 PER CENT CO_2 IS REQUIRED FOR PRIMARY ISOLATION OF:
A. Brucella suis
B. Brucella abortus
C. Both
D. Neither
Ref. 6 - p. 558

1754. X AND V FACTORS ARE REQUIRED FOR GROWTH BY:
A. Haemophilus aegyptius
B. Haemophilus influenzae
C. Both
D. Neither
Ref. 6 - p. 416

1755. HYPERSENSITIVITY TO DIPHTHERIA IS INDICATED BY A:
A. Positive Schick test
B. Positive Moloney test
C. Both
D. Neither
Ref. 6 - pp. 433, 435

1756. BILE SOLUBILITY:
A. Pneumococci
B. Streptococci
C. Both
D. Neither
Ref. 6 - p. 364

SECTION VII - GENERAL REVIEW
OF INFECTIOUS DISEASES

1757. RESISTANCE AGAINST GROUP A STREPTOCOCCI INCLUDES:
 A. Antibacterial immunity related to type specific M protein component
 B. Antitoxic immunity against erythrogenic toxin
 C. Both
 D. Neither   Ref. 5 - p. 381

1758. DESQUAMATION OF THE SKIN MAY FOLLOW:
 A. Rubella   C. Both
 B. Measles   D. Neither
                                          Ref. 4 - pp. 792, 806

1759. TEST OF SUSCEPTIBILITY TO SCARLET FEVER:
 A. Schultz-Charlton   C. Both
 B. Dick   D. Neither
                                          Ref. 6 - p. 380

1760. MULTINUCLEATED GIANT CELLS ARE FOUND IN SMEARS FROM VESICLES OF:
 A. Smallpox   C. Both
 B. Chickenpox   D. Neither
                                          Ref. 4 - p. 918

1761. MENINGITIS CAUSED BY N. MENINGITIDIS MAY BE:
 A. Epidemic   C. Both
 B. Endemic   D. Neither
                                          Ref. 6 - p. 404

1762. MENINGOCOCCAL BACTERIAL ENDOCARDITIS CAN BE CAUSED BY:
 A. Group A meningococci   C. Both
 B. Group B meningococci   D. Neither
                                          Ref. 6 - p. 406

1763. THE DISEASE HAS AN ACUTE ONSET:
 A. Infectious hepatitis   C. Both
 B. Serum hepatitis   D. Neither
                                          Ref. 4 - p. 967

1764. ROMAÑA'S SIGN:
 A. African trypanosomiasis   C. Both
 B. Chagas' disease   D. Neither
                                          Ref. 11 - p. 64

1765. THE ADULTS OF SCHISTOSOMA JAPONICUM ARE LOCATED IN THE VENULES OF THE:
 A. Small intestine   C. Both
 B. Large intestine   D. Neither
                                          Ref. 11 - p. 181

1766. TINEA CRURIS IS CAUSED BY:
 A. Epidermophyton floccosum   D. Both
 B. Trichophyton spp.   E. Neither
                                          Ref. 8 - p. 560

1767. THE USUAL PORTAL OF ENTRY FOR POLIOVIRUSES IS BELIEVED TO BE THROUGH:
 A. The nasal mucosa
 B. The alimentary tract
 C. Both
 D. Neither   Ref. 4 - p. 438

SECTION VII - GENERAL REVIEW OF INFECTIOUS DISEASES

1768. IN THE MAJORITY OF CASES, INFECTION WITH POLIOVIRUSES CAUSES:
- A. Invasion of the central nervous system and paralytic disease
- B. Fever and respiratory or gastrointestinal symptoms
- C. Both
- D. Neither

Ref. 4 - p. 446

1769. ECHOVIRUSES CAUSE:
- A. Aseptic meningitis
- B. Common colds
- C. Both
- D. Neither

Ref. 15 - p. 368

1770. HERPANGINA DUE TO COXSACKIE VIRUS INFECTION CAUSES LESIONS THAT ARE CHARACTERISTICALLY LOCATED ON THE:
- A. Anterior pillars of the fauces, uvula and palate
- B. Lips, gums, and anterior portion of the tongue
- C. Both
- D. Neither

Ref. 4 - p. 497

1771. MEASLES MAY BE TRANSMITTED BY:
- A. Droplet infection from a patient in the catarrhal stage
- B. Chronic human carriers
- C. Both
- D. Neither

Ref. 4 - p. 790

1772. KOPLIK'S SPOTS:
- A. Are present in nearly all cases of unmodified measles
- B. Persist for the duration of the exanthem of measles
- C. Both
- D. Neither

Ref. 4 - p. 792

1773. IN ROSEOLA INFANTUM:
- A. Koplik's spots are occasionally found
- B. The rash usually appears as the temperature returns to normal
- C. Both
- D. Neither

Ref. 4 - p. 811

1774. PERTUSSIS MAY BE FOLLOWED BY:
- A. Chronic bronchitis
- B. Bronchial asthma
- C. Both
- D. Neither

Ref. 6 - p. 425

1775. THE CONGENITAL AND NEONATAL HERPES INFECTIONS ARE PREDOMINANTLY CAUSED BY:
- A. H. hominis type 2
- B. H. hominis type 1
- C. Both
- D. Neither

Ref. 6 - p. 861

SECTION VII - GENERAL REVIEW OF INFECTIOUS DISEASES

AFTER EACH OF THE FOLLOWING CASE HISTORIES ARE SEVERAL MULTIPLE CHOICE QUESTIONS BASED ON THE HISTORY. ANSWER BY CHOOSING THE MOST APPROPRIATE ANSWER:

CASE HISTORY - Questions 1776-1780

An 8 year-old boy is seen at his home because of fever, sore throat, and vomiting. He was well until the early morning when he awoke complaining of sore throat. His temperature at that time was 102° F. He also complained of mild headache and vomited twice. His mother later noted that the glands in his neck were swollen and the doctor was called. Physical examination revealed a moderately ill child with a beefy red edematous pharynx and enlarged cervical lymph nodes. Temperature was 102.5°F.; pulse 130. The remainder of the physical examination is negative.

1776. THE MOST LIKELY DIAGNOSIS IS:
 A. Diphtheria
 B. Infectious mononucleosis
 C. Measles
 D. Streptococcal infection
 E. Vincent's angina   Ref. 5 - p. 377

Throat culture is taken and therapy with aspirin is begun. No antibiotics are given. The temperature is 101.5°F. Twelve hours later, the skin is warm and dry, and a generalized erythematous rash with punctate congestion of the skin papillae and hair follicles is noted.

1777. THE MOST LIKELY DIAGNOSIS IS NOW:
 A. Allergy to aspirin
 B. Infectious mononucleosis
 C. Measles
 D. Scarlet fever
 E. Rubella   Ref. 5 - p. 378

1778. THE FOLLOWING TEST COULD BE USED TO CONFIRM THE DIAGNOSIS OF SCARLET FEVER:
 A. Heterophile agglutination test
 B. The Schick test
 C. The Schultz-Charlton reaction
 D. The Dick test
 E. Culture of one of the skin lesions
          Ref. 6 - p. 380

1779. WHICH ONE IS NOT A COMPLICATION OR SEQUELA OF THIS DISEASE?:
 A. Peritonsillar abcess
 B. Otitis media
 C. Acute hemorrhagic glomerular nephritis
 D. Acute rheumatic fever
 E. Encephalomyelitis   Ref. 6 - p. 380

1780. TREATMENT SHOULD BE INSTITUTED WITH:
 A. Specific antitoxin
 B. Penicillin
 C. Sulfonamides
 D. Cortisone
 E. None of the above   Ref. 6 - p. 380

SECTION VII - GENERAL REVIEW OF INFECTIOUS DISEASES

CASE HISTORY - Questions 1781-1785

A 25 year-old pregnant woman is admitted to the maternity floor of a general hospital in early labor. Physical examination at that time reveals, in addition to the normal findings of pregnancy, a diffuse swelling of the right cheek anterior to the ear and extending below the angle of the jaw and under the mandible. There is a moderate tenderness. Temperature is 101°F. Her oldest child had a similar disease some time before the mother became ill.

1781. DIFFERENTIAL DIAGNOSIS SHOULD INCLUDE:
A. Bacterial infection of the parotid gland
B. Recurrent parotitis
C. Tumor of the parotid gland
D. Mumps parotitis
E. All of the above
Ref. 4 - p. 761

1782. THE WBC IS 6,500 WITH 60% LYMPHOCYTES. ASSUME THAT THE DIAGNOSIS IS MUMPS AND SHE WAS INFECTED BY THE OLDER CHILD. THE AVERAGE INCUBATION PERIOD IS:
A. One week
B. Two weeks
C. Three weeks
D. Four weeks
E. Five weeks
Ref. 4 - p. 759

1783. IF THE DIAGNOSIS IS INDEED MUMPS:
A. The newborn baby will posses strong immunity
B. The newborn baby is certain to be deformed
C. The virus could not have been transmitted through the placenta
D. The newborn baby should be immunized immediately
E. None of the above
Ref. 4 - p. 759

1784. WHICH OF THE FOLLOWING COMPLICATIONS IS LEAST LIKELY TO OCCUR WITH MUMPS?:
A. Orchitis
B. Pneumonitis
C. Pancreatitis
D. Meningo-encephalitis
E. Inflammation of the ovaries
Ref. 4 - p. 760

1785. AFTER THE MOTHER RECOVERS FROM MUMPS:
A. She will have a positive complement-fixation test
B. She will have a negative skin test
C. She will be susceptible to a repeat attack involving the other parotid
D. She is likely to be sterile
E. She may have an elevated sedimentation rate for several weeks
Ref. 4 - p. 758

SECTION VII - GENERAL REVIEW OF INFECTIOUS DISEASES

CASE HISTORY - Questions 1786-1790

An 8 year-old boy with a slight cold and temperature of 101°F. was taken to the office of a man who was not a physician complaining of headache and slight stiff neck. Local heat and manipulation of the cervical spine produced some relief. Ten grains of aspirin reduced the temperature to 100°F. in 3 hours. The child vomited once before going to bed that night. The following morning the temperature was 103°F. and the child was sleepy. His mother decided to permit him to stay in bed at his request. At noon he could not be aroused and a physician was called. He found a comatose boy with a rapid pulse, nuchal rigidity and a positive Kernig's sign.

1786. THE DIAGNOSTIC PROCEDURE MOST LIKELY TO AID IN THE CORRECT DIAGNOSIS IS A:
A. Complete blood count
B. Blood sugar determination
C. Urinalysis and BUN determination
D. Lumbar puncture
E. Chest X-ray and tuberculin test   Ref. 5 - p. 446

White blood count is 25,000 with 75% polymorphonuclear leukocytes. Blood sugar is 80 mg. per 100 ml. Urinalysis and BUN are normal. Spinal fluid is turbid. Chest X-ray reveals an old Ghon complex and tuberculin test is positive. Smears of the spinal fluid reveal gram-negative diplococci.

1787. THE ORGANISM IS MOST LIKELY :
A. M. tuberculosis
B. Meningococcus
C. Pneumococcus
D. Gonococcus
E. H. influenzae   Ref. 5 - p. 446

1788. FURTHER EXAMINATION OF THE SPINAL FLUID IS LIKELY TO SHOW:
A. A very high cell count
B. Increased protein
C. Low or no sugar content
D. Abnormal colloidal gold curve
E. All of the above   Ref. 5 - p. 446

1789. IF THIS WERE H. INFLUENZAE MENINGITIS, THE COMPLICATION LEAST LIKELY TO OCCUR IS:
A. Waterhouse-Friderichsen syndrome
B. Subdural effusion
C. Hydrocephalus
D. Otitis media
E. Pneumonia   Ref. 5 - p. 448

1790. IF THIS WERE MENINGOCOCCAL MENINGITIS, THE TREATMENT OF CHOICE IS:
A. Penicillin G
B. Chloromycetin
C. Sulfisoxazole (Gantrisin)
D. Tetracycline
E. Antimeningococcus serum   Ref. 15 - p. 175

SECTION VII - GENERAL REVIEW
OF INFECTIOUS DISEASES

CASE HISTORY - Questions 1791-1795

A 40 year-old man is admitted to the hospital with headache, vomiting, nuchal rigidity and temperature of 102°F. He was completely well until five days before when he developed upper respiratory symptoms and slight fever for 48 hours. He was then apparently well until the day of admission when the above symptoms began. He also complains of spontaneous muscle pain in the left lower extremity and slight difficulty swallowing. Physical examination reveals that he is irritable and has tachycardia. There is demonstrable weakness of some muscle groups in the left leg and absence of sensation over the anterior thigh and lower leg. There is an epidemic of poliomyelitis raging in the community. This man has never had poliomyelitis immunizations.

1791. IN POLIOMYELITIS:
 A. The entry of the virus is by the olfactory route
 B. The time interval between virus implantation and the "minor illness" phase is 2 weeks
 C. The time interval between exposure and the onset of symptoms of the CNS phase of disease is on the average 17 days
 D. Virus multiplication and excretion in individuals vaccinated with Salk vaccine occur within 24 hours
 E. Paralytic forms are most common in infants when epidemic strains have been introduced for the first time into remote areas
 Ref. 4 - p. 446

1792. THE MOST LIKELY TIME OF THE YEAR FOR POLIOMYELITIS TO OCCUR IN TEMPERATE CLIMATES:
 A. January
 B. March
 C. May
 D. August
 E. December
 Ref. 4 - p. 451

1793. EXAMINATION OF THE SPINAL FLUID IS LIKELY TO SHOW:
 A. Elevated cell count
 B. Rise and return to normal of protein
 C. Normal glucose concentration
 D. All of the above
 E. None of the above
 Ref. 4 - p. 449

1794. THE RELATIVE EPIDEMIOLOGIC IMPORTANCE OF CLINICALLY APPARENT AND INAPPARENT FORMS IN THE SPREAD OF POLIOMYELITIS IS DEPENDENT UPON:
 A. The degree of virulence of the infecting virus
 B. The virus type
 C. The age of the exposed susceptibles
 D. All of the above
 E. None of the above
 Ref. 4 - p. 452

1795. THE PATIENT'S DIFFICULTY IN SWALLOWING IS MOST LIKELY DUE TO:
 A. Spinal involvement
 B. Cerebral cortical involvement
 C. Bulbar involvement
 D. Coxsackie virus infection
 E. Foreign body in the esophagus
 Ref. 4 - p. 447

SECTION VII - GENERAL REVIEW OF INFECTIOUS DISEASES

CASE HISTORY - Questions 1796-1800

A 5 year-old boy is admitted to a large general hospital because of severe cough and vomiting of 7 days duration. The child was completely well until 3 weeks prior to the admission when he developed signs of an upper respiratory infection with rhinitis, fever and slight cough. The symptoms continued, but during the week prior to admission, the cough became more severe, paroxysmal and ended with an inspiratory crowing sound. The child would then vomit a small amount of food or mucus. There was no respiratory difficulty, except during and immediately after an episode of coughing. The child had not received any immunization during infancy and had received no therapy during the present illness. There was no history of exposure to contagious diseases.

1796. THE MOST LIKELY DIAGNOSIS, BASED ON THE HISTORY, IS:
A. Measles
B. Croup
C. Diphtheria
D. Foreign body in the trachea
E. Pertussis   Ref. 5 - p. 745

1797. ONE FACT AGAINST THE DIAGNOSIS OF MEASLES IS:
A. The presence of fever
B. The presence of a severe cough
C. The length of the prodromal period
D. The absence of history of exposure to measles
E. The absence of respiratory difficulty
                            Ref. 5 - p. 745

1798. IF THIS IS PERTUSSIS:
A. Transmission is by droplet infection
B. It is transmitted by a human carrier
C. Immunity developed will be short-lived
D. Passively transferred immunity from mother to the newborn is long-lasting
E. Highest mortality rate is in the 5-6 year age group
                            Ref. 5 - p. 744

1799. IF THIS IS PERTUSSIS, WHICH OF THE FOLLOWING WILL NOT BE OF ANY THERAPEUTIC VALUE?:
A. Human immune serum
B. Aureomycin
C. Pertussis vaccine
D. Chloromycetin
E. Terramycin   Ref. 5 - p. 747

1800. PERTUSSIS MUST BE DISTINGUISHED FROM:
A. Allergic bronchitis
B. Bronchopneumonia
C. Atypical viral pneumonia
D. Spasmodic coughs resulting from infected adenoids or sinuses
E. All of the above   Ref. 5 - p. 747

REFERENCES

1. Brown, H. W.: <u>Basic Clinical Parasitology</u>, 3rd Edition, Appleton-Century-Crofts, Inc., New York, N. Y., 1969
2. Burrows, S.: <u>Textbook of Microbiology</u>, 19th Edition, W. B. Saunders Co., Philadelphia, Pa., 1968
3. Frobisher, M.: <u>Fundamentals of Microbiology</u>, 8th Edition, W. B. Saunders Co., Philadelphia, Pa., 1968
4. Horsfall, F. L., Jr. and I. Tamm: <u>Viral and Rickettsial Infections of Man</u>, 4th Edition, J. B. Lippincott Co., Philadelphia, Pa., 1965
5. Dubos, R. J. and J. G. Hirsch: <u>Bacterial and Mycotic Infections of Man</u>, 4th Edition, J. B. Lippincott Co., Phildelphia, Pa., 1965
6. Joklik, W. K. and D. T. Smith: <u>Zinsser Microbiology</u>, 15th Edition, Appleton-Century-Crofts, Inc., New York, N. Y., 1972
7. Raffel, S.: <u>Immunity</u>, 2nd Edition, Appleton-Century-Crofts, Inc., New York, N. Y., 1961
8. Conant, N. F., D. T. Smith, R. D. Baker and J. L. Callaway: <u>Manual of Clinical Mycology</u>, 3rd Edition, W. B. Saunders Co., Philadelphia, Pa., 1971
9. Faust, E. C., P. C. Beaver and R. C. Jung: <u>Animal Agents and Vectors of Human Disease</u>, 3rd Edition, Lea and Febiger, Philadelphia, Pa., 1968
10. Boyd, W. C.: <u>Fundamentals of Immunology</u>, 4th Edition, John Wiley and Sons, Inc., New York, N. Y., 1966
11. Larsh, J. E., Jr.: <u>Outline of Medical Parasitology</u>, McGraw-Hill Book Co., Inc., New York, N. Y., 1964
12. Davis, B. D., R. Dulbecco, H. N. Eisen, H. S. Ginsberg and W. B. Wood, Jr.: <u>Microbiology</u>, Hoeber Medical Division, Harper & Row, Publ., Inc., New York, N. Y., 1967
13. Weiser, R. S., Q. N. Myrvik and N. N. Pearsall: <u>Fundamentals of Immunology</u>, Lea and Febiger, Philadelphia, Pa., 1970
14. Rhodes, A. J. and C. E. Van Rooyen: <u>Textbook of Virology</u>, 5th Edition, Williams & Wilkins Co., Baltimore, Md., 1968
15. Jawetz, E., J. L. Melnick and E. A. Adelberg: <u>Review of Medical Microbiology</u>, 10th Edition, Lange Medical Publ, Los Altos, Calif., 1972
16. Bellanti, J. A.: <u>Immunology</u>, W. B. Saunders Co., Philadelphia, Pa., 1971

ANSWER KEY

The authors have made every effort to thoroughly verify the questions and answers. However, in a volume of this size, some ambiguities and possible inaccuracies may appear; therefore, if in doubt, consult your references.

THE PUBLISHERS

1. E	52. A	103. B	154. D	205. E
2. A	53. D	104. D	155. C	206. C
3. D	54. D	105. A	156. C	207. C
4. B	55. C	106. C	157. A	208. B
5. B	56. A	107. C	158. B	209. B
6. E	57. D	108. C	159. C	210. E
7. D	58. B	109. B	160. D	211. B
8. B	59. A	110. D	161. C	212. B
9. C	60. B	111. A	162. C	213. B
10. C	61. B	112. C	163. B	214. A
11. D	62. D	113. C	164. D	215. C
12. A	63. C	114. C	165. C	216. A
13. A	64. D	115. C	166. D	217. E
14. B	65. C	116. C	167. A	218. A
15. C	66. C	117. A	168. C	219. E
16. C	67. D	118. C	169. C	220. D
17. C	68. C	119. A	170. B	221. B
18. C	69. B	120. A	171. E	222. C
19. E	70. A	121. C	172. B	223. C
20. B	71. D	122. D	173. C	224. A
21. B	72. C	123. A	174. D	225. C
22. E	73. A	124. C	175. B	226. C
23. E	74. C	125. E	176. D	227. C
24. B	75. C	126. A	177. A	228. D
25. C	76. D	127. D	178. C	229. C
26. A	77. C	128. C	179. D	230. C
27. A	78. C	129. A	180. D	231. C
28. D	79. C	130. D	181. D	232. C
29. C	80. F	131. D	182. B	233. D
30. B	81. D	132. D	183. A	234. C
31. B	82. B	133. A	184. C	235. B
32. C	83. A	134. C	185. D	236. C
33. B	84. D	135. D	186. B	237. C
34. D	85. D	136. C	187. C	238. D
35. A	86. A	137. C	188. E	239. C
36. T	87. A	138. B	189. D	240. D
37. F	88. B	139. C	190. C	241. A
38. T	89. C	140. D	191. A	242. D
39. T	90. E	141. C	192. A	243. C
40. F	91. C	142. E	193. A	244. B
41. T	92. B	143. E	194. E	245. B
42. T	93. D	144. B	195. D	246. D
43. F	94. B	145. A	196. B	247. C
44. T	95. B	146. B	197. D	248. D
45. T	96. E	147. D	198. B	249. D
46. C	97. D	148. E	199. D	250. C
47. D	98. A	149. E	200. B	251. D
48. A	99. E	150. B	201. D	252. E
49. D	100. A	151. A	202. E	253. E
50. A	101. A	152. C	203. D	254. D
51. C	102. D	153. C	204. D	255. C

ANSWER KEY

256. D	314. A	372. F	430. T	488. A
257. D	315. C	373. C	431. T	489. A
258. B	316. B	374. A	432. F	490. A
259. B	317. B	375. C	433. T	491. B
260. B	318. E	376. B	434. T	492. C
261. D	319. C	377. C	435. T	493. A
262. D	320. E	378. C	436. E	494. A
263. B	321. A	379. D	437. B	495. A
264. C	322. A	380. D	438. D	496. B
265. B	323. E	381. B	439. A	497. C
266. B	324. D	382. C	440. C	498. A
267. A	325. A	383. D	441. E	499. A
268. E	326. E	384. A	442. A	500. C
269. B	327. A	385. D	443. D	501. E
270. E	328. E	386. A	444. B	502. D
271. B	329. B	387. D	445. C	503. E
272. D	330. A	388. E	446. B	504. D
273. E	331. B	389. A	447. D	505. D
274. D	332. E	390. C	448. A	506. A
275. E	333. C	391. A	449. C	507. C
276. D	334. B	392. F	450. E	508. C
277. A	335. A	393. C	451. C	509. E
278. B	336. C	394. C	452. A	510. D
279. E	337. E	395. D	453. D	511. A
280. D	338. A	396. E	454. B	512. C
281. C	339. D	397. B	455. E	513. A
282. D	340. D	398. C	456. C	514. A
283. A	341. A	399. B	457. A	515. C
284. D	342. A	400. D	458. B	516. E
285. B	343. A	401. C	459. E	517. D
286. C	344. C	402. A	460. D	518. C
287. A	345. D	403. B	461. A	519. B
288. A	346. A	404. D	462. B	520. E
289. B	347. B	405. C	463. A	521. D
290. C	348. D	406. A	464. A	522. D
291. C	349. C	407. A	465. D	523. C
292. D	350. A	408. D	466. E	524. D
293. C	351. A	409. F	467. B	525. D
294. B	352. D	410. C	468. A	526. C
295. A	353. A	411. C	469. A	527. D
296. C	354. B	412. A	470. C	528. B
297. B	355. C	413. B	471. A	529. E
298. A	356. A	414. D	472. E	530. B
299. C	357. A	415. F	473. D	531. A
300. A	358. B	416. T	474. D	532. E
301. A	359. E	417. T	475. A	533. D
302. A	360. A	418. T	476. E	534. C
303. C	361. E	419. F	477. C	535. B
304. C	362. D	420. T	478. A	536. D
305. C	363. A	421. T	479. D	537. A
306. A	364. A	422. T	480. C	538. B
307. D	365. A	423. F	481. A	539. B
308. C	366. D	424. T	482. C	540. E
309. B	367. A	425. T	483. D	541. B
310. C	368. E	426. F	484. B	542. E
311. D	369. A	427. T	485. D	543. C
312. C	370. B	428. T	486. C	544. A
313. C	371. C	429. T	487. C	545. A

ANSWER KEY

546. A	604. A	662. 3-A	720. A	778. D
547. C	605. B	663. 4-A	721. C	779. D
548. E	606. D	664. 3-B	722. A	780. D
549. B	607. D	665. 2-C	723. C	781. E
550. C	608. D	666. A	724. D	782. B
551. D	609. D	667. C	725. A	783. E
552. A	610. E	668. E	726. C	784. C
553. D	611. A	669. B	727. A	785. A
554. D	612. A	670. D	728. C	786. B
555. B	613. E	671. E	729. A	787. C
556. B	614. B	672. B	730. D	788. C
557. E	615. A	673. A	731. A	789. A
558. C	616. A	674. D	732. C	790. D
559. C	617. B	675. C	733. C	791. D
560. C	618. D	676. B	734. A	792. C
561. B	619. D	677. B	735. A	793. B
562. B	620. B	678. E	736. D	794. E
563. D	621. B	679. D	737. A	795. C
564. D	622. A	680. A	738. D	796. D
565. A	623. C	681. E	739. C	797. E
566. B	624. C	682. C	740. D	798. B
567. A	625. E	683. A	741. D	799. C
568. B	626. A	684. D	742. A	800. D
569. B	627. D	685. B	743. C	801. B
570. A	628. B	686. B	744. E	802. A
571. A	629. B	687. A	745. D	803. E
572. B	630. B	688. D	746. B	804. D
573. A	631. B	689. E	747. D	805. C
574. B	632. A	690. C	748. A	806. E
575. B	633. E	691. B	749. B	807. A
576. C	634. A	692. D	750. C	808. E
577. B	635. A	693. A	751. C	809. A
578. A	636. 2-A	694. E	752. B	810. B
579. B	637. 1-C	695. C	753. A	811. B
580. C	638. 3-B	696. C	754. E	812. C
581. B	639. 5-A	697. E	755. D	813. A
582. A	640. 4-A	698. A	756. B	814. A
583. A	641. 3-B	699. D	757. C	815. D
584. A	642. 2-A	700. B	758. D	816. A
585. A	643. 4-B	701. E	759. A	817. D
586. B	644. 5-A	702. D	760. C	818. B
587. B	645. 4-C	703. A	761. E	819. C
588. A	646. 3-B	704. B	762. C	820. D
589. A	647. 3-C	705. C	763. D	821. D
590. A	648. 3-A	706. B	764. E	822. C
591. B	649. 2-A	707. E	765. E	823. A
592. A	650. 5-C	708. A	766. C	824. B
593. A	651. 5-B	709. D	767. E	825. D
594. A	652. 2-A	710. C	768. D	826. D
595. A	653. 2-B	711. C	769. D	827. B
596. A	654. 1-A	712. D	770. E	828. E
597. B	655. 5-A	713. E	771. D	829. E
598. D	656. 5-B	714. A	772. B	830. A
599. A	657. 1-C	715. B	773. E	831. E
600. B	658. 3-A	716. C	774. A	832. B
601. C	659. 5-C	717. B	775. D	833. D
602. E	660. 4-A	718. C	776. B	834. C
603. C	661. 5-B	719. C	777. C	835. D

ANSWER KEY

836. B	894. 1-A	952. E	1010. C	1068. B
837. C	895. 5-B	953. B	1011. D	1069. A
838. A	896. 4-C	954. A	1012. D	1070. B
839. A	897. 3-A	955. C	1013. C	1071. C
840. C	898. 2-A	956. D	1014. A	1072. A
841. B	899. 2-A	957. A	1015. C	1073. B
842. C	900. 5-A	958. D	1016. A	1074. D
843. E	901. E	959. A	1017. E	1075. B
844. E	902. D	960. D	1018. A	1076. A
845. B	903. A	961. C	1019. E	1077. C
846. C	904. B	962. A	1020. C	1078. A
847. D	905. C	963. B	1021. D	1079. A
848. C	906. B	964. B	1022. B	1080. C
849. E	907. E	965. A	1023. D	1081. C
850. A	908. D	966. A	1024. A	1082. A
851. D	909. A	967. E	1025. B	1083. C
852. B	910. C	968. D	1026. E	1084. C
853. C	911. B	969. C	1027. A	1085. C
854. E	912. E	970. D	1028. C	1086. T
855. B	913. E	971. B	1029. D	1087. T
856. E	914. C	972. D	1030. A	1088. F
857. C	915. D	973. B	1031. A	1089. T
858. E	916. A	974. D	1032. C	1090. T
859. E	917. A	975. C	1033. C	1091. T
860. D	918. D	976. C	1034. A	1092. F
861. D	919. A	977. D	1035. C	1093. F
862. D	920. B	978. D	1036. E	1094. T
863. E	921. B	979. D	1037. A	1095. T
864. A	922. A	980. D	1038. B	1096. F
865. C	923. E	981. E	1039. A	1097. T
866. C	924. E	982. B	1040. B	1098. T
867. D	925. D	983. E	1041. A	1099. T
868. D	926. D	984. C	1042. A	1100. F
869. C	927. C	985. D	1043. C	1101. T
870. E	928. A	986. A	1044. B	1102. T
871. D	929. A	987. B	1045. E	1103. F
872. C	930. D	988. A	1046. B	1104. T
873. B	931. C	989. E	1047. C	1105. F
874. C	932. D	990. C	1048. B	1106. A
875. B	933. D	991. C	1049. E	1107. C
876. 5-B	934. A	992. D	1050. E	1108. B
877. 5-A	935. E	993. C	1051. D	1109. B
878. 2-B	936. E	994. C	1052. A	1110. B
879. 5-A	937. D	995. C	1053. E	1111. A
880. 4-B	938. A	996. C	1054. A	1112. A
881. 2-B	939. B	997. C	1055. A	1113. B
882. 3-B	940. A	998. B	1056. A	1114. B
883. 5-C	941. A	999. B	1057. B	1115. C
884. 1-A	942. A	1000. E	1058. B	1116. A
885. 3-B	943. B	1001. B	1059. A	1117. A
886. 4-C	944. B	1002. A	1060. D	1118. B
887. 5-A	945. C	1003. D	1061. C	1119. B
888. 4-A	946. D	1004. E	1062. C	1120. C
889. 5-A	947. D	1005. A	1063. A	1121. A
890. 4-A	948. C	1006. D	1064. C	1122. A
891. 4-C	949. C	1007. A	1065. B	1123. C
892. 5-C	950. C	1008. A	1066. D	1124. C
893. 5-C	951. B	1009. C	1067. C	1125. C

ANSWER KEY

1126. C	1184. B	1242. C	1300. D	1358. A
1127. C	1185. C	1243. D	1301. E	1359. B
1128. A	1186. C	1244. C	1302. D	1360. A
1129. C	1187. D	1245. D	1303. E	1361. A
1130. A	1188. A	1246. D	1304. B	1362. C
1131. E	1189. E	1247. D	1305. A	1363. A
1132. B	1190. B	1248. B	1306. C	1364. C
1133. D	1191. 5-B	1249. A	1307. B	1365. A
1134. A	1192. 4-A	1250. D	1308. C	1366. A
1135. C	1193. 1-A	1251. F	1309. D	1367. C
1136. C	1194. 1-B	1252. D	1310. A	1368. A
1137. E	1195. 4-A	1253. C	1311. D	1369. B
1138. A	1196. 5-A	1254. A	1312. E	1370. C
1139. B	1197. 4-A	1255. B	1313. E	1371. A
1140. D	1198. 3-C	1256. D	1314. B	1372. A
1141. C	1199. 3-A	1257. A	1315. D	1373. C
1142. E	1200. 1-B	1258. F	1316. E	1374. A
1143. B	1201. 1-B	1259. C	1317. C	1375. B
1144. A	1202. 4-B	1260. C	1318. B	1376. B
1145. D	1203. 2-A	1261. D	1319. D	1377. C
1146. B	1204. 4-A	1262. D	1320. E	1378. A
1147. E	1205. 5-B	1263. C	1321. C	1379. A
1148. D	1206. 4-B	1264. A	1322. D	1380. A
1149. A	1207. 3-B	1265. D	1323. C	1381. C
1150. C	1208. 3-B	1266. C	1324. C	1382. D
1151. B	1209. 5-C	1267. B	1325. A	1383. A
1152. C	1210. 4-C	1268. C	1326. A	1384. A
1153. A	1211. 5-A	1269. D	1327. E	1385. B
1154. E	1212. 3-A	1270. C	1328. D	1386. B
1155. D	1213. 4-C	1271. D	1329. C	1387. C
1156. B	1214. 2-A	1272. C	1330. B	1388. A
1157. D	1215. 1-C	1273. C	1331. B	1389. A
1158. A	1216. 3-C	1274. D	1332. E	1390. E
1159. E	1217. 4-B	1275. C	1333. E	1391. E
1160. C	1218. 2-C	1276. D	1334. D	1392. B
1161. B	1219. 2-A	1277. C	1335. B	1393. A
1162. C	1220. 1-C	1278. E	1336. A	1394. E
1163. A	1221. 3-A	1279. E	1337. C	1395. A
1164. E	1222. 4-C	1280. D	1338. E	1396. A
1165. D	1223. 1-B	1281. C	1339. C	1397. B
1166. B	1224. 4-A	1282. B	1340. B	1398. D
1167. D	1225. 4-A	1283. D	1341. D	1399. C
1168. A	1226. D	1284. A	1342. C	1400. A
1169. E	1227. D	1285. B	1343. D	1401. E
1170. C	1228. D	1286. A	1344. D	1402. F
1171. C	1229. F	1287. B	1345. B	1403. C
1172. D	1230. D	1288. A	1346. B	1404. D
1173. B	1231. C	1289. B	1347. E	1405. B
1174. A	1232. D	1290. D	1348. B	1406. 5-B
1175. E	1233. E	1291. D	1349. C	1407. 2-B
1176. B	1234. C	1292. A	1350. D	1408. 5-A
1177. D	1235. C	1293. B	1351. D	1409. 2-C
1178. E	1236. A	1294. B	1352. C	1410. 5-A
1179. A	1237. B	1295. D	1353. C	1411. 1-A
1180. C	1238. C	1296. D	1354. E	1412. 4-A
1181. D	1239. F	1297. A	1355. E	1413. 3-C
1182. E	1240. C	1298. C	1356. A	1414. 4-B
1183. A	1241. F	1299. C	1357. A	1415. 5-A

1416. D	1474. D	1532. C	1590. C	1648. E
1417. C	1475. E	1533. A	1591. A	1649. B
1418. A	1476. A	1534. B	1592. C	1650. A
1419. B	1477. D	1535. A	1593. B	1651. C
1420. E	1478. C	1536. B	1594. A	1652. C
1421. D	1479. D	1537. E	1595. C	1653. D
1422. E	1480. B	1538. D	1596. A	1654. B
1423. A	1481. E	1539. C	1597. B	1655. A
1424. B	1482. A	1540. A	1598. A	1656. E
1425. C	1483. B	1541. B	1599. A	1657. C
1426. E	1484. D	1542. D	1600. B	1658. B
1427. D	1485. A	1543. A	1601. E-3	1659. B
1428. A	1486. D	1544. A	1602. A-2	1660. B
1429. C	1487. C	1545. A	1603. E-1	1661. D
1430. B	1488. B	1546. B	1604. C-1	1662. E
1431. C	1489. D	1547. B	1605. E-1	1663. C
1432. E	1490. B	1548. A	1606. D-1	1664. A
1433. B	1491. B	1549. C	1607. A-2	1665. D
1434. A	1492. D	1550. D	1608. A-2	1666. D
1435. D	1493. C	1551. A	1609. D-1	1667. C
1436. B	1494. C	1552. A	1610. B-1	1668. E
1437. D	1495. B	1553. B	1611. E-1	1669. D
1438. E	1496. E	1554. D	1612. D-2	1670. B
1439. C	1497. B	1555. B	1613. C-2	1671. B
1440. A	1498. A	1556. C	1614. A-2	1672. A
1441. A	1499. A	1557. A	1615. D-2	1673. D
1442. A	1500. E	1558. A	1616. B	1674. E
1443. E	1501. A	1559. A	1617. D	1675. B
1444. C	1502. A	1560. E	1618. A	1676. C
1445. E	1503. B	1561. A	1619. F	1677. A
1446. B	1504. C	1562. E	1620. E	1678. D
1447. D	1505. C	1563. A	1621. D	1679. C
1448. A	1506. B	1564. E	1622. B	1680. F
1449. D	1507. C	1565. D	1623. F	1681. A
1450. E	1508. C	1566. C	1624. C	1682. C
1451. B	1509. C	1567. B	1625. A	1683. A
1452. A	1510. A	1568. C	1626. F	1684. C
1453. E	1511. D	1569. B	1627. B	1685. D
1454. A	1512. C	1570. B	1628. D	1686. A
1455. B	1513. C	1571. B	1629. A	1687. B
1456. E	1514. A	1572. A	1630. C	1688. C
1457. E	1515. C	1573. C	1631. A	1689. A
1458. C	1516. A	1574. A	1632. B	1690. D
1459. E	1517. A	1575. B	1633. C	1691. A
1460. C	1518. C	1576. A	1634. E	1692. D
1461. C	1519. C	1577. B	1635. D	1693. C
1462. E	1520. C	1578. A	1636. B	1694. C
1463. D	1521. C	1579. B	1637. A	1695. D
1464. B	1522. B	1580. B	1638. E	1696. F
1465. D	1523. C	1581. A	1639. C	1697. B
1466. E	1524. C	1582. B	1640. D	1698. B
1467. C	1525. B	1583. A	1641. A	1699. B
1468. C	1526. C	1584. B	1642. B	1700. C
1469. C	1527. D	1585. B	1643. C	1701. C
1470. E	1528. A	1586. A	1644. D	1702. A
1471. B	1529. B	1587. A	1645. D	1703. B
1472. B	1530. A	1588. B	1646. D	1704. E
1473. E	1531. C	1589. A	1647. A	1705. D

ANSWER KEY

1706. C
1707. E
1708. A
1709. D
1710. B
1711. D
1712. C
1713. A
1714. E
1715. B
1716. C
1717. D
1718. A
1719. B
1720. E
1721. C
1722. A
1723. E
1724. B
1725. D
1726. E
1727. A
1728. B
1729. C
1730. D
1731. A
1732. A
1733. B
1734. D
1735. B
1736. E
1737. D
1738. A
1739. A
1740. A
1741. A
1742. B
1743. B
1744. A
1745. B
1746. D
1747. B
1748. B
1749. C
1750. C
1751. A
1752. B
1753. B
1754. C
1755. D
1756. A
1757. C
1758. C
1759. B
1760. B
1761. C
1762. C
1763. A

1764. B
1765. A
1766. C
1767. B
1768. D
1769. C
1770. A
1771. A
1772. A
1773. B
1774. C
1775. A
1776. D
1777. D
1778. C
1779. E
1780. B
1781. E
1782. C
1783. E
1784. B
1785. A
1786. D
1787. B
1788. E
1789. A
1790. A
1791. C
1792. D
1793. D
1794. D
1795. C
1796. E
1797. C
1798. A
1799. C
1800. E

**AVAILABLE AT YOUR LOCAL BOOKSTORE
OR USE THIS ORDER FORM**

MEDICAL EXAMINATION PUBLISHING CO., INC.
65-36 Fresh Meadow Lane, Flushing, N.Y. 11365

Date: _____

Please send me the following books:

_____

_____

_____

_____

_____

☐ Payment enclosed to save postage.

☐ Bill me. I will remit payment within 30 days.

Name _____

Address _____

City & State _____ Zip _____

(Please print)

**AVAILABLE AT YOUR LOCAL BOOKSTORE
OR USE THIS ORDER FORM**

MEDICAL EXAMINATION PUBLISHING CO., INC.
65-36 Fresh Meadow Lane, Flushing, N.Y. 11365

Date: _____

Please send me the following books:

_____

_____

_____

_____

_____

☐ Payment enclosed to save postage.

☐ Bill me. I will remit payment within 30 days.

Name _____

Address _____

City & State _____ Zip _____

(Please print)

**AVAILABLE AT YOUR LOCAL BOOKSTORE
OR USE THIS ORDER FORM**

MEDICAL EXAMINATION PUBLISHING CO., INC.
65-36 Fresh Meadow Lane, Flushing, N.Y. 11365

Date: _____

Please send me the following books:

_____
_____
_____
_____
_____

☐ Payment enclosed to save postage.

☐ Bill me. I will remit payment within 30 days.

Name _____

Address _____

City & State _____ Zip _____

(Please print)

OTHER BOOKS AVAILABLE

Quan.	ITEMS	Code	Unit Price	Quan.	ITEMS	Code	Unit Price
	MEDICAL EXAM REVIEW BOOKS				**SPEC. BOARD REV. BKS.** *(Cont'd.)*		
	Vol. 1 Comprehensive	101	$12.00		Family Practice Specialty Board Review	309	$10.00
	Vol. 2 Clinical Medicine	102	7.00		Internal Medicine Specialty Board Review	303	10.00
	Vol. 2A Txtbk. Study Guide of Int. Med.	123	7.00		Neurology Specialty Board Review	306	10.00
	Vol. 2B Txtbk. Study Guide of Int. Med.	130	7.00		Obstetrics-Gynecology Spec. Bd. Review	304	10.00
	Vol. 3 Basic Sciences	103	7.00		Pathology Specialty Board Review	305	10.00
	Vol. 4 Obstetrics-Gynecology	104	7.00		Pediatrics Specialty Board Review	301	10.00
	Vol. 4A Textbk. Study Guide of Gynecology	152	7.00		Physical Medicine Spec. Bd. Review	308	10.00
	Vol. 5 Surgery	105	7.00		Surgery Specialty Board Review	302	10.00
	Vol. 5A Textbk. Study Guide of Surgery	150	7.00		The Psychiatry Boards	307	8.00
	Vol. 6 Public Health & Prev. Medicine	106	7.00		**STATE BOARD REVIEW BOOKS**		
	Vol. 8 Psychiatry & Neurology	108	7.00		Medical State Board Exam. Rev. - Part 1	411	9.00
	Vol. 11 Pediatrics	111	7.00		Medical State Board Exam. Rev. - Part 2	412	9.00
	Vol. 12 Anesthesiology	112	7.00		Cardiopulmonary Techn. Exam. Rev. - Vol. 1	473	7.00
	Vol. 13 Orthopaedics	113	10.00		Dental Exam. Review Book - Vol. 1	431	7.50
	Vol. 14 Urology	114	10.00		Dental Exam. Review Book - Vol. 2	432	7.50
	Vol. 15 Ophthalmology	115	10.00		Dental Exam. Review Book - Vol. 3	433	7.50
	Vol. 16 Otolaryngology	116	10.00		Dental Hygiene Exam. Review - Vol. 1	461	7.00
	Vol. 17 Radiology	117	10.00		Emergency Med. Techn. Exam. Rev. - Vol. 1	465	7.00
	Vol. 18 Thoracic Surgery	118	10.00		Emergency Med. Techn. Exam. Rev. - Vol. 2	466	7.00
	Vol. 19 Neurological Surgery	119	15.00		Immunology Exam. Review Book - Vol. 1	424	7.00
	Vol. 20 Physical Medicine	128	10.00		Inhalation Therapy Exam. Review - Vol. 1	471	7.00
	Vol. 21 Dermatology	127	10.00		Inhalation Therapy Exam. Review - Vol. 2	344	7.00
	Vol. 22 Gastroenterology	141	10.00		Laboratory Asst. Exam. Rev. Bk. - Vol. 1	455	7.00
	Vol. 23 Child Psychiatry	126	10.00		Medical Librarian Exam. Rev. Bk. - Vol. 1	495	7.00
	Vol. 24 Pulmonary Diseases	143	10.00		Medical Record Library Science - Vol. 1	496	7.00
	ECFMG Exam Review - Part One	120	7.00		Medical Techn. Exam. Review - Vol. 1	451	7.00
	ECFMG Exam Review - Part Two	121	7.00		Medical Techn. Exam. Review - Vol. 2	452	7.00
	BASIC SCIENCE REVIEW BOOKS				Occupational Therapy Exam. Rev. - Vol. 1	475	7.00
	Anatomy Review	201	6.00		Pharmacy Exam. Review Book - Vol. 1	421	7.00
	Biochemistry Review	202	6.00		Physical Therapy Exam. Review - Vol. 1	481	7.00
	Heart & Vascular Systems Basic Sciences	212	7.00		Physical Therapy Exam. Review - Vol. 2	482	7.00
	Microbiology Review	203	6.00		X-Ray Technology Exam. Rev. - Vol. 1	441	7.00
	Nervous System Basic Sciences	210	7.00		X-Ray Technology Exam. Rev. - Vol. 2	442	7.00
	Pathology Review	204	6.00		**NURSING EXAM REVIEW BOOKS**		
	Pharmacology Review	205	6.00		Vol. 1 Medical-Surgical Nursing	501	4.00
	Physiology Review	206	6.00		Vol. 2 Psychiatric-Mental Health Nursing	502	4.00
	Respiratory System Basic Sciences	213	7.00		Vol. 3 Maternal-Child Health Nursing	503	4.00
	Anatomy Textbook Study Guide	124	7.00		Vol. 4 Basic Sciences	504	4.00
	Histology Textbook Study Guide	151	7.00		Vol. 5 Anatomy and Physiology	505	4.00
	Medical Physiology Textbk. Study Guide	155	7.00		Vol. 6 Pharmacology	506	4.00
	SPECIALTY BOARD REVIEW BOOKS				Vol. 7 Microbiology	507	4.00
	Dermatology Specialty Board Review	311	10.00		*(Continued on reverse side)*		

Prices subject to change

OTHER BOOKS AVAILABLE

Quan.	ITEMS	Code	Unit Price	Quan.	ITEMS	Code	Unit Price
	Vol. 8 Nutrition & Diet Therapy	508	$ 4.00		J. ART. COMPILATIONS (Cont'd.)		
	Vol. 9 Community Health	509	4.00		Immunosuppressive Therapy Journal Art.	526	$20.00
	Vol. 10 History & Law of Nursing	510	4.00		Institutional Laundry Journal Articles	789	8.00
	Vol. 11 Fundamentals of Nursing	511	4.00		Lithium & Psychiatry Journal Articles	520	15.00
	Practical Nursing Examination Rev. - Vol. 1	711	4.50		Outpatient Services Journal Articles	794	8.00
	MEDICAL OUTLINE SERIES				Psychosomatic Medicine Current J. Art.	788	12.00
	Cancer Chemotherapy	631	10.00		Selected Papers in Inhalation Therapy	523	10.00
	Otolaryngology	661	8.00		**TYPIST HANDBOOKS**		
	Psychiatry	621	8.00		Medical Typist's Guide for Hx & Phys.	976	4.50
	Urology	611	8.00		Radiology Typist Handbook	981	4.50
	CASE STUDY BOOKS				Surgical Typist Handbook	991	4.50
	Cutaneous Medicine Case Studies	014	7.00		**OTHER BOOKS**		
	ECG Case Studies	003	7.00		Acid Base Homeostasis	601	4.00
	Endocrinology Case Studies	008	10.00		Allergy Annual Review	325	12.00
	Gastroenterology Case Studies	004	10.00		Bailey & Love's Short Practice of Surgery	900	20.00
	Hematology Case Studies	020	10.00		Benign & Malignant Bladder Tumors	932	15.00
	Infectious Diseases Case Studies	011	7.00		Blood Groups	860	2.50
	Neurology Case Studies	006	10.00		Clinical Diagnostic Pearls	730	4.50
	Otolaryngology Case Studies	021	10.00		Concentrations of Solutions	602	3.00
	Pediatric Hematology Case Studies	018	10.00		Critical Care Manual	983	10.00
	Respiratory Care Case Studies	019	7.00		Cryogenics in Surgery	754	24.00
	Urology Case Studies	017	10.00		Diagnosis & Treatment of Breast Lesions	748	15.00
	SELF-ASSESSMENT BOOKS				English-Spanish Guide for Med. Personnel	721	2.50
	Self-Assess. Cur. Knldge - Nurse Anesthet.	715	7.00		Fundamental Orthopedics	603	4.50
	Self-Assess. Cur. Knldge - Infect. Diseases	263	10.00		Guide to Medical Reports	962	4.50
	Self-Assess. Cur. Knldge in Neurology	254	10.00		Handbook of Medical Emergencies	635	7.00
	Self-Assess. Cur. Knldge in Pathology	253	10.00		Human Anatomical Terminology	982	3.00
	Self-Assess. Cur. Knldge in Pediatrics	256	10.00		Illustrated Laboratory Techniques	919	10.00
	Self-Assess. Cur. Knldge in Psychiatry	252	10.00		Introduction to Blood Banking	975	8.00
	Self-Assess. Cur. Knldge in Rheumatology	258	10.00		Introduction to the Clinical History	729	3.00
	Self-Assess. Cur. Knldge in Surgery	250	10.00		Multilingual Guide for Medical Personnel	961	2.50
	S.A.C.K. Surgery for Family Physicians	259	10.00		Neoplasms of the Gastrointestinal Tract	736	20.00
	Self-Assess. Cur. Knldge in Urology	251	10.00		Neurology Handbook	604	8.00
	JOURNAL ARTICLE COMPILATIONS				Neurophysiology Study Guide	600	6.00
	Ambulance Service Journal Articles	517	10.00		Ophthalmology Rev. Essay Quest. & Ans.	347	10.00
	Blood Banking & Immunohemat. Jour. Art.	798	10.00		Outpatient Hemorrhoidectomy Lig. Tech.	752	12.50
	Emergency Room Journal Articles	795	8.00		Practical Points in Gastroenterology	733	7.00
	Hodgkin's Disease Journal Articles	515	12.00		Profiles in Surgery, Gynec. & Obstetrics	963	5.00
	Hosp. & Inst. Eng. & Maintenance J. Art.	793	8.00		Radiological Physics Exam. Review	486	10.00
	Hosp. Electronic Data Process. J. Art.	791	8.00		Skin, Heredity & Malignant Neoplasms	744	20.00
	Hosp. Housekeeping Journal Articles	790	8.00		Testicular Tumors	743	20.00
	Hosp. Pharmacy Journal Articles	799	10.00		Tissue Adhesives in Surgery	756	24.00
	Hosp. Security & Safety Journal Articles	796	8.00				
	Human Cytomegalovirus Journal Articles	522	15.00				

Prices subject to change